Lecture Notes in Earth Sciences   80

Editors:
S. Bhattacharji, Brooklyn
G. M. Friedman, Brooklyn and Troy
H. J. Neugebauer, Bonn
A. Seilacher, Tuebingen and Yale

W0245784

Springer-Verlag Berlin Heidelberg GmbH

Serge A. Shapiro   Peter Hubral

Elastic Waves
in Random Media

Fundamentals of
Seismic Stratigraphic Filtering

With 63 Figures and 4 Tables

 Springer

Authors

Prof. Dr. Serge A. Shapiro
Ecole Nationale Supérior de Géologie
Institut National Polytechnique de Lorraine
CRPG-CNRS, Bat. G
Rue du D.-M. Roubault
BP 40, F-54501 Vandoeuvre les Nancy, France
E-mail: shapiro@ensg.u-nancy.fr

Prof. Dr. Peter Hubral
Wave Inversion Technology Group
Geophysical Institute, University of Karlsruhe
Hertzstr. 16, D-76187 Karlsruhe, Germany
E-mail: peter.hubral@physik.uni-karlsruhe.de

"For all Lecture Notes in Earth Sciences published till now please see final pages of the book"

Cataloging-in-Publication data applied for

Die Deutsche Bibliothek - CIP-Einheitsaufnahme

Shapiro, Sergei A.:
Elastic waves in random media : fundamentals of seismic
stratigraphic filtering / Serge A. Shapiro ; Peter Hubral. - Berlin ;
Heidelberg ; New York ; Barcelona ; Budapest ; Hong Kong ;
London ; Milan ; Paris ; Singapore ; Tokyo : Springer, 1999
  (Lecture notes in earth sciences ; 80)

ISSN 0930-0317
ISBN 978-3-540-65006-5   ISBN 978-3-540-49775-2 (eBook)
DOI 10.1007/978-3-540-49775-2

This work is subject to copyright. All rights are reserved, whether the whole or part of the material is concerned, specifically the rights of translation, reprinting, re-use of illustrations, recitation, broadcasting, reproduction on microfilms or in any other way, and storage in data banks. Duplication of this publication or parts thereof is permitted only under the provisions of the German Copyright Law of September 9, 1965, in its current version, and permission for use must always be obtained from Springer-Verlag. Violations are liable for prosecution under the German Copyright Law.

© Springer-Verlag Berlin Heidelberg 1999
Originally published by Springer-Verlag Berlin Heidelberg New York in 1999.

The use of general descriptive names, registered names, trademarks, etc. in this publication does not imply, even in the absence of a specific statement, that such names are exempt from the relevant protective laws and regulations and therefore free for general use.

Typesetting: Camera ready by authors
SPIN: 10693554   32/3142-543210 - Printed on acid-free paper

This book treats various generalizations of the classical O'Doherty-Anstey formula in order to describe stratigraphic filtering effects. These are the effects that can be observed when acoustic, elastic and electromagnetic waves propagate through a stack of thin layers. Our aim was to treat this topic in a comprehensive manner and present compact results in a didactically simple way, emphasizing the physics of the wave-propagation phenomena. We do not claim mathematical rigor in all our derivations, however, we are pleased to have obtained quite simple generalizations of the original O'Doherty-Anstey formula for various useful situations. Hence, we are confident that many of the results presented in this book are just as valuable as the original classical result of O'Doherty and Anstey and will be considered for equally many practical applications.

Preface

Stacks of thin layers are a physical reality in any sedimentary basin. Typical sonic logs (as well as shear-wave and density logs) show that sedimentary rocks are composed of layers with thicknesses of the order of $0.1m$ to $10m$. Sometimes a micro layering of even smaller characteristic sizes can be observed. This means that to a reservoir at $4km$ depth the wave field must transmit through hundreds or thousands of layers. *The influence of thinly layered structures on transmitted seismic wavefields is called stratigraphic filtering.* An understanding and a reliable description of this effect is of importance for seismic exploration and seismology. Of course, there never exists a perfectly parallel layered geological structure. However, in many cases this model, also called 1-D model (note that all media under consideration are assumed to be heterogeneous along one spatial axis only) can be used as a first and useful approximation to describe the stratigraphic filtering.

On the other hand, multilayered stacks can be observed not only in the solid Earth. Oceans, atmosphere, artificial composite materials and some biological media very often have similar structures. Moreover, many real systems are 1-D or quasi 1-D just due to their physical nature (like e.g., communication lines). Thus, the problems of stratigraphic filtering are of more general character than seismological only.

The attention of seismologists and exploration geophysicists has long been attracted to the problem of seismic wave propagation in parallel layered 1-D elastic media. There exists a number of books, which are entirely or partly devoted to seismic waves in layered structures (Aki and Richards, 1980; Robinson and Treitel, 1980; Brekhovskikh and Godin, 1989; Kennett, 1983; Tygel and Hubral, 1987, etc.), without however, giving attention to the diverse aspects of stratigraphic filtering as done here.

Often an exact computation of the recorded wavefield (i.e., the synthetic seismogram) in multilayered media is unwieldy and in fact not necessary. Moreover, the exact information on the physical parameters of these layers is never known. The principal task of this book is to provide a simple description of mean properties (like the attenuation coefficient and the phase velocity) of the transmitted wavefield in terms of a very restricted number of statistical properties of the medium (e.g., the correlation distance and the variances of the density and velocity-log fluctuations). These statistics can be expected to be more reliable than the details in the log measurements, which are always influenced by the drilling- and logging techniques.

The so-called generalized O'Doherty-Anstey formulas, which are a central topic in this book, fulfill the above mentioned task. In contrast to deterministic models and approaches discussed in the above mentioned books, here the seismic wave propagation in sediments is considered as a problem of wave propagation in randomly multilayered 1-D media. Effects of non perfect 1-D multilayering are not considered. The attention will mainly be concentrated on properties of plane wavefields normally or obliquely transmitted through a multilayered 1-D medium (i.e., transmissivity). An approximation of wavefields being reflected from such a medium (i.e., reflectivity) will be also discussed. Effects of inelasticity will be touched in our consideration of poroelastic media and in Section 8.6. This book is organized as follows:

In Chapter 1 we give a short description of the main assumptions, concepts and results of this book and a historical introduction to the subject.

In Chapter 2 we review the basic and most important concepts of the theory of random processes as well as some important statistical and physical terms related to localization, random media, averaging and self-averaged quantities.

In Chapter 3 we discuss normal-incidence plane-wave propagation in the most simple and in seismic processing widely used 1-D model of sediments, i.e. the Goupillaud model (Goupillaud, 1961) - where the stratification is represented by homogeneous layers equidistantly spaced in traveltime. A plane wavefield propagates in such systems as a single mode. In spite of its very schematic character the Goupillaud model is of great practical importance. Important signal processing steps, like e.g., deconvolution have been designed for this model. It provides a good possibility to understand the behavior of not only the main part of the transmitted wavefield following immediately its first arrival, but its coda also. Additionally, a very clear description of the reflectivity is possible in the frame of this model.

In Chapter 4 we continue to consider wavefield propagation in the single-mode regime. Now, however, the oblique incidence in general 1-D inhomogeneous structures is under consideration. We obtain the attenuation coefficients and vertical phase increments (i.e., real parts of vertical components of the wave vectors) for transmissivities in the case of scalar waves. These can be pressure waves in a fluid or SH-polarized shear waves in a solid. The corresponding derivation is performed in a small-perturbation approximation. This is valid up to the second order in the fluctuations of the density and elastic modulus, i.e., the validity domain of the O'Doherty-Anstey approximation. One can view the result as being a second-order Rytov approximation.

In Chapter 5 we consider multi-mode wavefields. The case of compression- (P) or shear- (SV) waves (i.e., vector wavefields) obliquely propagating in a 1-D heterogeneous solid will be studied. Relatively simple explicit formulas are obtained for the attenuation coefficients and phase increments of the transmitted waves. Finally, in this chapter, an approach is presented which is based on the invariant-embedding method. This provides an elegant and rather general way of solution for the reflected as well as transmitted wavefields.

In Chapter 6 the solutions obtained for scalar and vector wavefields in elastic media are used to discuss the seismic time-harmonic transmissivity in more depth. Its kinematic and dynamic properties like phase velocities and attenuation coefficients are analyzed. Such effects like frequency-dependent anisotropy, shear-wave splitting and angle-dependent attenuation are described.

In Chapter 7 these results are used to derive the transmissivity in the time domain (i.e., the transient transmissivity). In this chapter we also give a number of numerical illustrations to better understand the theory.

In Chapter 8 we discuss practical applications of the theory of stratigraphic filtering for amplitude corrections in the amplitude-variation with offset (AVO) analysis.

In Chapter 9 we consider the multi-mode propagation of the wavefield in poroelastic 1-D inhomogeneous media. We describe there an important mechanism of the attenuation of seismic waves called the inter-layer flow. In addition, we approach the question of seismic signatures of the permeability.

In Chapter 10 a short discussion of the reflectivity of elastic waves finalizes our consideration.

In this book, which is dedicated to a very specialized topic of seismic-wave-propagation theory, we must assume that the reader possesses certain backgrounds of wave propagation, statistics, communication theory, etc. Depending on the individual background, she or he would find access to the whole book by not necessarily starting with Chapter 1. One can start selecting the chapters with the subject one is most familiar with and proceed from these to the less familiar topics in the other chapters.

In connection with writing this book we have used results obtained in a close cooperation with many of our colleagues, who have worked at the Geophysical Institute of the Karlsruhe University at different times. We would like to express our gratitude to them for an excellent cooperation and the stimulating discussions we had with them. These colleagues are Holger Zien, Jörg Schleicher, Stephan Gelinsky, Martin Widmaier, Thilo Müller, Olaf Knot, Tobias Müller, Uli Werner, Boris Gurevich, Sven Treitel, Bjorn Ursin, Martin Tygel and Norbert Gold.

The results reported here were in a large part obtained during the second phase of the German-Norwegian geoscientific cooperation: a project financially supported by the BMBF (Germany) and STATOIL (Norway). An additional support in the year 1997 was received from the DFG (Germany; Contract SH 55/1-1).

Table of Contents

1 Introduction 1

   1.1 Problems and Results 1

   1.2 Short Historical Review 6

2 Random Media and Wave Propagation 9

   2.1 Statistical Moments of Random Processes: Basic Concepts . . . 9

   2.2 Statistical Properties of Sediments 12

   2.3 Averaging 13

   2.4 Localization 14

   2.5 Self-Averaged Quantities 16

   2.6 Products of Random Matrices 18

3 Normal-Incidence Waves in a Stack of Layers 23

   3.1 Recursive Description of the Wavefield 24

   3.2 Exact Expressions for the Wavefield 26

   3.3 Explicit Approximations for the Transmissivity 29

   3.4 Generalized Primary and Coda 32

   3.5 Appendix: Approximation of the Polynomial A_n 40

4 Oblique Incidence of Scalar Waves 45

   4.1 Model and Dynamic Equations 45

   4.2 Strategy of the Solution 48

   4.3 Attenuation and Phase of the Transmissivity 49

   4.4 Results for Scalar Waves 52

4.5 Results in Terms of Reflection-Coefficient-Series Spectra 54

4.6 Appendix: Details of the Derivation for ψ and γ 57

5 Elastic P-SV Waves 61

5.1 Model and Dynamic Equations 62

5.2 Transformation of the Dynamic Equations 64

5.3 Invariant Embedding 67

5.4 Transmissivity in Case of Small Fluctuations 70

5.5 Transmissivity in Stationary Random Media 74

    Vertical incidence 76

    Constant density and shear velocity 76

    General reduction to the acoustic case 76

5.6 Transmissivity for Non-Stationary Random Media 77

5.7 Validity Conditions of the Solution 80

5.8 Appendix: Explicit Expressions for Some Quantities. 82

5.9 Appendix: Alternative Derivation of Equations (5.35)-(5.37) . . . 83

6 Frequency-Dependent Properties of Stratigraphic Filtering . 87

6.1 Low-Frequency Asymptotic Solution 88

6.2 High-Frequency Asymptotic Solution 89

6.3 Whole-Frequency Domain Solution 91

6.4 Frequency-Dependent Shear-Wave Splitting 98

6.5 Fluctuations of the Attenuation Coefficient 103

6.6 Inversion for Statistics of Stratifications 105

7 Transient Transmissivity 109

   7.1 Coherent and Incoherent Parts 109

   7.2 Numerical Transmissivity and Pulse Stabilization 113

8 Stratigraphic Filtering and Amplitude Variation with Offset   119

   8.1 Correction Method 120

   8.2 Well-Log Data 126

   8.3 Amplitude Processing 133

   8.4 Estimation of Statistical Macro Models from Well Logs 137

   8.5 Correction of AVO Responses 142

   8.6 Scattering or Intrinsic Absorption? 145

9 Stratigraphic Filtering in Poroelastic Media 151

   9.1 Model and Dynamic Equations 152

   9.2 Time-Harmonic Transmissivity 155

   9.3 Below the Critical Frequency. 157

   9.4 Attenuation and Permeability 161

   9.5 Appendix: Periodic Media – Low-Frequency Limit 167

   9.6 Appendix: Random Media – High-Frequency Limit 168

10 Reflectivities of Multilayered Structures 169

   10.1 Normal-Incidence Plane Wave 169

   10.2 Oblique-Incidence Elastic Plane Waves 173

   10.3 Dynamic-Equivalent-Medium Reflectivity 175

   10.4 Intensity of the Reflectivity 177

11 Instead of Conclusions 181

References 183

Index 189

1 Introduction

We start this book with a description of its philosophy, terminology, aims and main results. A short historical review will be given in the second section of this Introduction.

1.1 Problems and Results

In this book we consider the transmission and reflection of acoustic/elastic waves in 1-D heterogeneous media. We call such media layered, multilayered or stratified. Sometimes we call them also stratified structures, stratified layers, stack of thin layers or laminations. In all these cases we consider spaces, half-spaces or layers, which are homogeneous along two coordinate axes and heterogeneous along one coordinate axis only. When we say 'thin layers' we mean that the thicknesses of elementary layers are much smaller then traveldistances under consideration. The same is implied when speaking about small-scale heterogeneities. Under large-scale heterogeneities we understand the heterogeneities on the scale of traveldistances.

Usually we will consider the following configuration: a stratified layer is embedded between two homogeneous half-spaces. A plane wave is incident from the first half-space (commonly the upper half-space) onto the multilayered structure. The influence of this structure on the plane wave transmitted into the second half-space is called *stratigraphic filtering*.

Indeed, if we work in the linear approximation of the elasticity theory (which is usually well justified in Seismology as well as in Non-Destructive Testing) the interaction of waves with heterogeneities can be described in terms of the linear-system theory as linear filtering. For instance, the employment of Green's functions and Fourier transforms to the problems of wave scattering and propagation can be conceived in terms of such linear filtering.

Let us consider a time-harmonic plane wave W_0 incident from the upper half-space onto a heterogeneous structure:

$$W_0 = \exp(-i\omega t + ik_z z + i\omega px), \tag{1.1}$$

where $i = \sqrt{-1}$, t is time, ω is the angular frequency, k_z the vertical component of the wavevector and p is the horizontal slowness (i.e., ωp is the horizontal component of the wavevector). The geometry of the problem is shown in Figure 1 . If we put the receiver below the bottom of the heterogeneous structure we will be able to measure an outgoing time-harmonic plane wave T propagating in the lower half-space. We call this wave *time-harmonic transmissivity* or, simply, *transmissivity*. Then, there exists a linear relationship between the input W_0 and the output T:

$$\bar{T} = S_f W_0, \tag{1.2}$$

where the quantity S_f can be called the *spatio-temporal spectrum of the stratigraphic filter*. It is interesting to remark that the time-harmonic transmissivity T is the frequency-wavevector characteristic of the stratigraphic filter (note, that the input plane wave W_0 is a single time- as well as spatial-harmonic component). Finally, the integral

$$\frac{1}{2\pi} \int_{-\infty}^{\infty} T d\omega \tag{1.3}$$

is called the *transient transmissivity*. It is the response of the stratigraphic filter for an input in the form of an obliquely incident delta-pulse plane wave.

In the following chapters we consider elastic and pressure waves. Thus, we work with the displacement vector and the pressure, respectively. Time-dependent distributions of the displacement and pressure (or other physical quantities, like the stress tensor) are called *wavefields*. Instead of the word 'wave' we use sometimes the word 'wavefield'. For example, for the geometry shown in Figure 1 in the case of time-harmonic pressure waves in a fluid, the wavefield in the lower half-space is the transmitted plane pressure wave. It is the same as the time-harmonic transmissivity. We can also call it the *transmitted wavefield*. In the considered case the wavefield in the uppermost half-space is an interference between the incident and reflected plane waves or between the incident and reflected wave-fields. It is clear also that the wavefield in the heterogeneous part of the medium

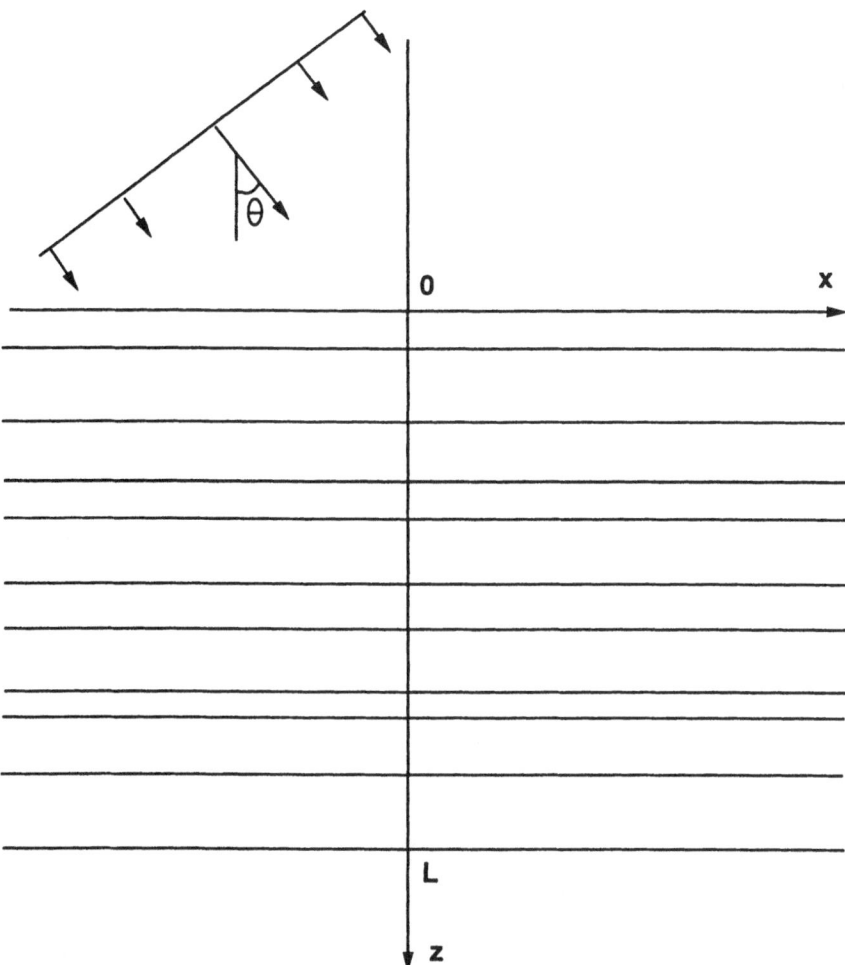

Fig. 1. Geometry of the problem: a plane wave is obliquely incident on a 1-D inhomogeneous structure. The transmitted and reflected wavefield is the subject of our study.

is an interference of many reflected and transmitted plane waves. The reflected plane wave will be also called *reflectivity*.

The main aim of this book is to describe the influence of multilayered structures on the transmitted seismic wavefields. The central results are the formulas for the time-harmonic transmissivities of the plane elastic P- and SV-waves (5.35)-(5.37). These formulas express the angle- and frequency-dependent amplitudes and phases of the transmitted waves in terms of simple statistical properties of sonic- and density logs: their auto- and cross-correlation functions. Formulas

(5.36)-(5.37) provide us with the attenuation coefficients and the phase velocities of the transmitted plane waves. These attenuation coefficients and phase velocities are frequency dependent. Because we obtain them for the total frequency range we say that these attenuation coefficients and phase velocities characterize the plane-wave propagation in homogeneous *dynamic-equivalent media*, which replace the actual stratified structures. Due to historical reasons we call these formulas the *generalized O'Doherty-Anstey formulas*. The corresponding formulas for the case of pressure waves in a stratified fluid and SH-waves in a stratified solid (4.18)-(4.22) are derived in Chapter 4. In some simple situations (normal incidence or scalar waves) the generalized O'Doherty-Anstey formulas have elegant forms if expressed in terms of Fourier transforms of reflection-coefficient series. These are the formulas (3.29), (3.31), (3.32), (3.52), (3.53) and (4.28)-(4.30).

The generalized O'Doherty-Anstey formulas are necessary to analyze all important properties of the stratigraphic filtering like the frequency-dependent velocity anisotropy, angle- and frequency-dependent attenuation and frequency-dependent shear-wave splitting. All these properties are analyzed and numerically illustrated in Chapter 6. In this chapter we give also explicit expressions for the attenuation coefficients and the vertical components of the real parts of the wavevectors (we call them *vertical phase increments*) for the P- and S-waves transmitted through a randomly-layered exponentially correlated structure (formulas 6.7). This particular case of laminations (or a good approximation of it) can be often observed in reality. Additionally, we discuss in Chapter 6 some methods of the transmissivity inversion for the statistical parameters of stratified structures.

It is important to note that formulas (5.36)-(5.37) are valid for the case, where the statistics of the log measurements is depth-independent (i.e., stationary). In Section 5.6 we show that these formulas are also valid in the case of depth-dependent averages of the P- and S-wave velocities and the density (i.e., nonstationary media). In this case a velocity/density macro model of the large-scale heterogeneous background medium must be extracted from sonic and density logs and the auto- and crosscorrelation functions of the small-scale fluctuations must be estimated. We call a set of such functions the *statistical macro model* of the small-scale fluctuations. This statistical macro model can be substituted in formulas (5.36)-(5.37) in order to estimate the effects of the stratigraphic filtering. In Chapter 8 we show how to apply this methodology in order to correct for the angle-dependent amplitude effect of the stratigraphic filtering when performing the AVO analysis. In this chapter we work with incident spherical waves and show the applicability of the generalized O'Doherty-Anstey formulas. This means that one can combine the ray-theoretical formalism with the generalized O'Doherty-Anstey formulas in order to perform a true-amplitude imaging or inversion.

Properties of the time-harmonic transmissivity of normally incident plane compressional waves in such complex systems like multilayered poroelastic saturated media are considered in Chapter 9. In this case the attenuation coefficient and the vertical phase increment are given by formulas (9.9)-(9.10). In the case of exponentially correlated laminations these formulas give the attenuation coefficients and the phase velocity expressed by formulas (9.11)-(9.12). These formulas show that in poroelastic media in addition to the elastic scattering there exists another attenuation mechanism important in the seismic frequency range (see formulas (9.12) and (9.20)). This is the *inter-layer flow*, which is the mechanism of inelastic absorption of seismic energy. This part of the seismic attenuation is permeability dependent, and therefore, it is of great interest for reservoir characterization. The analysis of this attenuation shows, however, that in heterogeneous media one must distinguish between the permeability, which controls the seismic attenuation and the permeability, which describes the fluid flow through multilayered systems.

The frequency-domain integration of the time-harmonic transmissivity, as given in formula (1.3), provides the transient transmissivity. In the case of elastic P- and S-waves the transient transmissivity is given by formula (7.2). In the case of exponentially correlated laminations this formula takes a more specific form taking formula (7.6) into account. Numerical simulations (see Chapters 3 and 7) show that the transient transmissivity based on a statistical macro model of laminations only describes the main part of the transmitted wavefield, which is close to the first arrivals. The reason of this is that the generalized O'Doherty-Anstey formulas have not a statistical but rather a deterministic origin. Indeed, the basic formulas (5.28), describing the logarithms of the elastic-wave time-harmonic transmissivity, are valid for any deterministic form of medium parameterization. The same follows from a detailed analysis of the time-harmonic and transient transmissivity of normally incident elastic waves given in Chapter 3. For instance, the time-harmonic transmissivity formulas (3.29), (3.31) and (3.32) have a completely deterministic character. From them not only the main part of the transient transmissivity can be obtained but also its coda. Moreover, formulas (3.29) and (3.31) provide an improvement of the O'Doherty-Anstey approximation. They also show that in order to describe statistically the coda of the transient transmissivity fourth statistical moments of the reflection coefficient series, corresponding to the heterogeneous medium, are necessary (see formulas (3.43) and (3.46)).

The time-harmonic transmissivity given by the generalized O'Doherty-Anstey approximation can be used to formulate the reflectivity. For instance, in the case of normal incidence we give deterministic and hybrid (statistic/deterministic) formulas of the reflectivity taking into account effects of multiple scattering. These are formulas (10.2)-(10.3). In the case of obliquely incident P- and S-waves, formulas (10.6)-(10.7) give the first-order (Born-type) and second-order reflectivity approximations. These approximations can be further improved by incorporating into them the dynamic-equivalent-medium wavevectors computed with the help of the generalized O'Doherty-Anstey approximation (as discussed in Section 10.3).

1.2   Short Historical Review

The principal kinematic and dynamic properties of seismic waves in horizontally layered media can be studied by different numerical methods as, for instance, the reflectivity method (Fuchs and Müller, 1971) for a point source or, the Thomson-Haskell (propagator matrix) method for a plane wave. Also, there are more computer-intensive methods available like finite-difference methods or the Alekseev-Mikhailenko method (see e.g., Korn, 1985; Kerner, 1990). However, in the case of many thin layers any numerical algorithm becomes very time consuming and does not help very much to clarify the physics of wave propagation.

Analytical methods are most suited to investigate multiple scattering as the main mechanism determining the wavefield properties in 1-D structures. Two very interesting and well-known results are published in the classical papers of Backus (1962) and O'Doherty and Anstey (1971). Backus averaged the elastic modulus directly in the wave equation. He thus obtained a description of a stack of isotropic layers in the form of an effectively anisotropic medium. Later Backus' theory, now called the *Backus averaging*, was generalized by Schoenberg and Muir (1989) to the case of a stack of anisotropic layers. Their description, however, is also valid for low-frequency wavefields only.

O'Doherty and Anstey (1972) provided, as a result of mainly heuristic reasoning, a formula for the time-harmonic transmissivity (see formulas (3.32) and (4.28)), which is valid in the whole-frequency domain. However, their formula describes the normally incident scalar plane waves only.

As a result of the practical importance of the O'Doherty-Anstey formula and of its originally incomplete derivation, various authors tried to improve its derivation. Banik et al. (1985) and Görich and Müller (1987) obtained the same expression for a statistically-averaged traveltime-corrected wavefield. Statistical averaging requires an ensemble of realizations of a random medium. Such an ensemble, however, never exists in reality. Resnick et al. (1986), using a deterministic derivation based on the *invariant-embedding method*, reduced the problem to a Riccati equation and avoided the problem of ensemble averaging. These authors, however, restricted their consideration to vertical incidence only. Ursin (1987) then simplified this derivation and generalized it to the case of oblique incidence of a plane pressure wave.

The invariant-embedding method has been in great depth developed and described for scalar waves vertically propagating in randomly layered media (Klyatskin and Saichev, 1992; Klyatskin, 1986; Klyatskin, 1980). For scalar waves,

Ursin (1987) and Jeffryes (1993) also used a deterministic derivation based on the invariant-embedding method. An analysis based on the invariant-embedding method has been performed by Ash et al., (1991) and by Lewicki and Papanicolaou (1994) for the scalar-wave reflectivity. Ursin (1983) proposed a scheme to apply the invariant-embedding method to vector waves. Also Norris (1995) provided a formulation of the invariant-embedding method for a very general system of first-order linear differential equations. In this book the invariant-embedding method also will be used to obtain solutions for the transmitted and reflected elastic vector wavefields.

A time-domain formulation of the multiple-scattering problem for an obliquely incident plane wave in an elastic 1-D medium was presented in a series of papers (Burridge et al. 1988; Burridge and Chang 1989; de Hoop et al. 1991a; de Hoop et al. 1991b; a mathematically very sound review of this approach can be found in Asch et al. 1991). These authors apply an averaging by smoothing rapidly varying functional terms in the perturbation series representing the transmissivity. As a result they obtain a solution in the form of an integro-differential equation. Later a solution for a viscoelastic 1-D medium was obtained by only applying a small perturbation approximation without any averaging (Burridge et al. 1993). This solution - which might even give an approximation of the coda of the transmissivity - requires the whole sonic-, shear- and density logs and still needs intensive numerical calculations. However, no explicit simple expressions for frequency-dependent kinematic and dynamic properties of the transmitted wavefield were formulated.

The physical foundations of the alternative (frequency-domain) approach described here are the phenomena of *wavefield localization* and *self-averaging* in randomly layered media. Localization controls the attenuation (and, therefore, owing to causality, also the phase and the velocity dispersion) of the seismic transmissivity due to multiple scattering. White et al. (1990) and Sheng et al. (1990) were the first who investigated the significance of localization for the propagation of scalar acoustic waves in sediments. Self-averaging, on the other hand, controls the statistics of self-averaged quantities of the wavefield. Both these fundamental physical phenomena were mathematically analyzed to a great depth in the now classical book of Lifshits et al. (1988).

The mathematical roots of self-averaging and localization go to properties of products of *random matrices* (Crisanti et al. 1993), which characterize corresponding properties of products of propagator matrices for wavefields in 1-D randomly layered structures. A series of theorems closely related to the self-averaging property of products of random matrices was proved by Lewicki and coauthors (Lewicki, 1994, Lewicki et al. 1994 and Lewicki et al 1996). An analysis of the localization effects for P- and SV elastic waves based on the random-matrix formalism and mainly concentrated to the low-frequency approximation was given by Kohler et al. (1996a; 1996b).

In a series of our own publications (Gelinsky and Shapiro, 1997; Shapiro and Zien, 1993; Shapiro et al., 1994a; Shapiro et al., 1994b; Shapiro and Hubral, 1994; Shapiro and Hubral, 1995; Shapiro et al, 1996; Shapiro and Hubral, 1996, Widmaier et al, 1996 and Werner and Shapiro, 1998) we tried to further develop the localization/self-averaging based approach and to extend it to elastic and poroelastic media. We concentrated our attention on the small-fluctuation approximations providing results in the total frequency range. This allowed us to describe such important properties of the stratigraphic filtering as the angle-dependent attenuation, frequency-dependent velocity anisotropy and shear-wave splitting. We also tried to show ways for practical applications of the gained description of stratigraphic filtering in seismic imaging and inversion. These results are summarized in this book.

2 Random Media and Wave Propagation

Seismic waves propagating in randomly multilayered media are subjected to stratigraphic filtering, whose physical reason is the multiple scattering by 1-D inhomogeneities. The direct consequence of multiple scattering is the wave localization, which appears approximately as an exponential attenuation of transmitted waves. The corresponding attenuation coefficients have complicated angular and frequency dependencies. Two more direct consequences of the multiple scattering (causally connected to the first one) are the velocity dispersion and anisotropy (transverse isotropy). The latter two effects are also very closely mutually related because the multiple scattering takes place in a stratified medium.

In this chapter the basic - and for all subsequent considerations - most important results on elastic waves in multilayered media as well as some statistical and physical concepts are reviewed. We give here introductory explanations for the terms averaging, localization and self-averaged quantities, which are used in the following chapters.

The effects of localization and self-averaging can be mathematically considered as consequences of asymptotic properties of products of random matrices. This has been demonstrated in a series of special theorems. We start this chapter, however, with a short summary of statistical principles, which are relevant for our consideration. Detailed introductions to the theory of random fields and statistic analysis can be found in many excellent books, e.g., Bendat and Piersol (1984). Here we follow the definitions given by Rytov et al, (1989a,b).

2.1 Statistical Moments of Random Processes: Basic Concepts

Let us consider a *random function* $f(z)$ of the depth z. We also call this function a *random process*. The random function f depends on a single spatial variable, which is the depth z. In order to completely define a random process we must provide all its $k-$point, or k-variate ($k = 1, 2, ...$) *probability densities* p_k. A function $p_k(z_1, f_1; z_2, f_2; ... z_k, f_k)$ for any number k of points $z_1, z_2, ..., z_k$ defines the probability

$$p_k df_1 df_2 ... df_k \tag{2.1}$$

that all the following k inequalities are valid:

$$f_1 \leq f(z_1) < f_1 + df_1, f_2 \leq f(z_2) < f_2 + df_2, ..., f_k \leq f(z_k) < f_k + df_k. \tag{2.2}$$

The non-negative functions p_k must obey special properties of symmetry, hierarchy and normalization (see Rytov et al., 1989a p. 84-86.) If all the k-point probability density functions are independent of any shift of the form $z_1, z_2, ..., z_k$ $\rightarrow z_1 + \delta z, z_2 + \delta z, ..., z_k + \delta z$, i.e., if for any function p_k

$$p_k(z_1, f_1; z_2, f_2; ... z_k, f_k) = p_k(z_1 + \delta z, f_1; z_2 + \delta z, f_2; ... z_k + \delta z, f_k), \tag{2.3}$$

then we will call $f(z)$ a *statistically homogeneous* or a *stationary process*. If only p_1 and p_2 are independent of any shift, then $f(z)$ is called *stationary-in-the-wide-sense*.

We see now that in order to describe a *random medium* (in terms of a single or many random processes corresponding to the physical properties of this medium, e.g., density, elastic moduli, porosity, etc.) we need an infinite number of probability-density functions. However, in order to describe the influence of the medium fluctuations on transmitted seismic wavefields (i.e., to describe the effects of stratigraphic filtering) usually a restricted number of functions p_k is enough. Moreover, usually only some first *statistical moments* are required.

The first statistical moment of the random process $f(z)$ at any depth z_1 is:

$$\langle f(z_1) \rangle = \int_{-\infty}^{\infty} f_1 p_1(z_1, f_1) df_1. \tag{2.4}$$

This quantity is also called the *mean value* or *mathematical expectation* of the process $f(z)$ at the depth z_1. If the random process is stationary in the wide sense, then the first statistical moment is z-independent, i.e., it is a constant.

The first statistical moment can be used in order to extract the *fluctuation* part of the random process, $\xi_f(z)$:

$$\xi_f(z) = f(z) - \langle f(z) \rangle. \tag{2.5}$$

This we will consider from now on rather then the random process $f(z)$ itself.

The second statistical moment of the process $f(z)$ is

$$
\langle \xi_f(z_1)\xi_f(z_2)\rangle = \int_{-\infty}^{\infty}\int_{-\infty}^{\infty}(f_1 - \langle f(z_1)\rangle)\ (\ f_2 - \langle f(z_2)\rangle) \times
$$
$$
\times\ p_2(z_1, f_1; z_2, f_2)df_1 df_2.
$$

(2.6)

We call this moment the *autocorrelation function*. Generally, the autocorrelation function is a function of two spatial positions. The quantity $\sigma_{ff}^2(z) = \langle \xi_f(z)^2\rangle$ is called the *variance* of the random process $f(z)$. The *standard deviation* of $\xi(z)$ is the positive square root of the variance. In the following it will be denoted as $\sigma_{ff}(z)$.

In the case of stationary random processes in-the-wide-sense the autocorrelation function is, however, the function of a single argument, the so-called *correlation lag $z = z_2 - z_1$*:

$$
\langle \xi_f(z_1)\xi_f(z_2)\rangle = B_{ff}(z_2 - z_1).
$$

(2.7)

Clearly, for such processes $\sigma_{ff}^2(z) = B_{ff}(0)$.

In the following we will consider mainly stationary random processes in-the-wide sense. For the sake of simplicity we will call them simply *stationary*. Later we will see that many properties of wavefields in random media are controlled by the Fourier spectrum of the autocorrelation function $B_{ff}(z)$. This is called the *power spectrum* (or *fluctuation spectrum*) of the process $f(z)$ and is given by:

$$
\hat{B}_{ff}(k) = \int_{-\infty}^{\infty}B_{ff}(z)e^{ikz}dz, \quad B_{ff}(z) = \frac{1}{2\pi}\int_{-\infty}^{\infty}\hat{B}_{ff}(k)e^{-ikz}dk,
$$

(2.8)

where k is the Fourier-domain wavenumber. It is clear that power spectra of random processes are real non-negative quantities (because of the well-known property of the Fourier transforms of autocorrelation functions, see e.g., Bath, 1974).

2.2   Statistical Properties of Sediments

Statistical properties of sediments can be studied by analyzing available bore-hole measurements, like, e.g., sonic logs or density logs (in Section 8.4 we give some examples of such an analysis). On the large scale (in the order of 1km), borehole measurements show significant non-stationarity. Therefore, usually, the physical properties of sediments are statistically analyzed after removal of long-wavelength trends and their mean values. Additional complications are caused by the influence which log tools and measurement methods have on the data. Such measurement effects can be considered as a kind of averaging of the real physical parameters. Therefore, they are disturbing the high-frequency compo-nents of power spectra. For a detailed consideration of this problem we refer the reader to the paper of Hsu and Burridge (1991).

We give here some examples of autocorrelation functions and power spectra of random processes often used for modeling or approximations of log mea-surements. The most simple autocorrelation functions are the exponential and Gaussian autocorrelation functions. In the following we call media with such sta-tistical properties simply *exponential* and *Gaussian*, respectively. In a Gaussian medium the autocorrelation function $B_{ff}(z)$ has the following form:

$$B_{ff}(z) = \sigma_{ff}^2 e^{-z^2/l^2} , \tag{2.9}$$

where l is the *correlation length* of the fluctuations and σ_{ff} is their standard deviation. The Fourier transform of this autocorrelation function is

$$\hat{B}_{ff}(k) = \sigma_{ff}^2 \sqrt{\pi} l e^{-k^2 l^2/4} . \tag{2.10}$$

The corresponding quantities of an exponential medium are

$$B_{ff}(z) = \sigma_{ff}^2 e^{-|z|/l} , \tag{2.11}$$

$$\hat{B}_{ff}(k) = 2\sigma_{ff}^2 \frac{l}{(1 + k^2 l^2)} . \tag{2.12}$$

If a medium is characterized by an exponential autocorrelation function, then the function expressing the depth dependence of the corresponding physical parame-ter will be non-differentiable (see e.g., an excellent exposition given by Chernov, 1960, p.9-10). Therefore, this physical parameter is discontinuously distributed in depth. This is typical for sedimentary stratifications. A more general example of discontinues structures are media with the von Karman autocorrelation function:

$$B_{ff}(z) = \sigma_{ff}^2 2^{1-\nu}\Gamma^{-1}(\nu)(|z|/l)^\nu K_\nu(|z|/l), \tag{2.13}$$

where K_ν is the modified Bessel function of the third kind (MacDonald function). The corresponding fluctuation spectrum has the following form:

$$\hat{B}_{ff}(k) = 2\sigma_{ff}^2\sqrt{\pi}\Gamma(\nu+1/2)\Gamma^{-1}(\nu)l(1+l^2k^2)^{-1/2-\nu}. \tag{2.14}$$

In the case $\nu = 1/2$ the von Karman function transforms into an exponential one. In general, the von Karman autocorrelation function is used to characterize what is called a turbulence fluctuations. Curves of depth dependencies of corresponding physical parameters of such media are fractals with the fractal dimension D

$$D = 2 - \nu, \quad 0 \leq \nu < 1. \tag{2.15}$$

Thus, an exponential autocorrelation function indicates a fractal behavior of the log measurements with the dimension $D = 1.5$.

2.3 Averaging

As done by Banik et al. (1985), also in this book *ensemble averaging* is used, however, for a different purpose. Ensemble averaging strictly implies that a set (i.e., an ensemble) of different media is supposed to exist a priori. This ensemble is called a *random medium*. Its elements are the *realizations of the random medium*. In order to obtain an ensemble average of any wavefield attribute (as e.g. the attenuation coefficient) this attribute has to be computed for each realization and then averaged over all the realizations of the random medium.

In the context of this book ensemble averaging is used only as a tool to simplify certain theoretical derivations and to obtain compact results. It is physically meaningless to average seismograms recorded in different sedimentary environments. As we have only one earth we have only one realization to be considered. This realization is assumed to be a "typical" one belonging to a random *ergodic* medium. Ergodic means that ensemble averaging can be replaced by *spatial averaging* in a single realization (i.e., depth averaging in the case of 1-D het-erogeneous sediments). All ergodic random processes are stationary. The crucial problem is linking the results obtained by ensemble averaging to those obtained for the one and only realization that we have for the earth. A linking is possible. It can be achieved by considering self-averaged quantities of the one and only wavefield that exists for the one and only layer stack which we have at our dis-posal. However, before addressing two fundamental self-averaged quantities we explain another important term, namely, *localization*.

2.4   Localization

Theoretical considerations related to localization were initiated in the classi-cal paper of Anderson (1958). Waves propagating in one- or two-dimensional random media show localization phenomena due to multiple scattering (Sheng, 1990; Sheng, 1995). Even in three dimensions, localization can be observed in some particular cases. Interference of the multiple backward scattered wavefield (which is always partially constructive) inevitably leads in 1-D random media to a concentration of the wavefield energy within exponentially small depth in-tervals close to plane-wave sources (Gredeskul and Freïlikher, 1990). This effect, if considered for a layered medium of thickness L (without intrinsic attenua-tion, i.e., without inelasticity), provides essentially an exponentially attenuated transmissivity. The characteristic scale that determines the attenuation is the frequency-dependent *localization length*, i.e., the reciprocal value of the attenu-ation coefficient.

In case L is large enough (L should be much larger than the localization length) the attenuation coefficient of the transmissivity will be L-independent and non-random, i.e., equal in each typical realization of the random medium. In this situ-ation, which is called *strong localization*, the wavefield has been already strongly decreased and its energy does not practically propagate downwards anymore.

For instance, in the case of a pressure wave in a fluid, localization implies that there exists a positive limit

$$\gamma = -\lim_{L\to\infty}\left\{\frac{1}{L}\ln\left|\frac{p_{true}(L)}{p_{true}(0)}\right|\right\} = -\lim_{L\to\infty}\left\{\frac{1}{L}\ln\frac{r(L)}{r(0)}\right\}, \qquad (2.16)$$

where $r(z)$ is the spatial envelope of the time-harmonic plane pressure wavefield $p_{true}(z)$, which is incident at the interface $z = 0$ on the layered medium of the thickness L. Figure 1 shows the geometry of the problem. A synthetic sonic log for a randomly layered stack is shown in Figure 2.

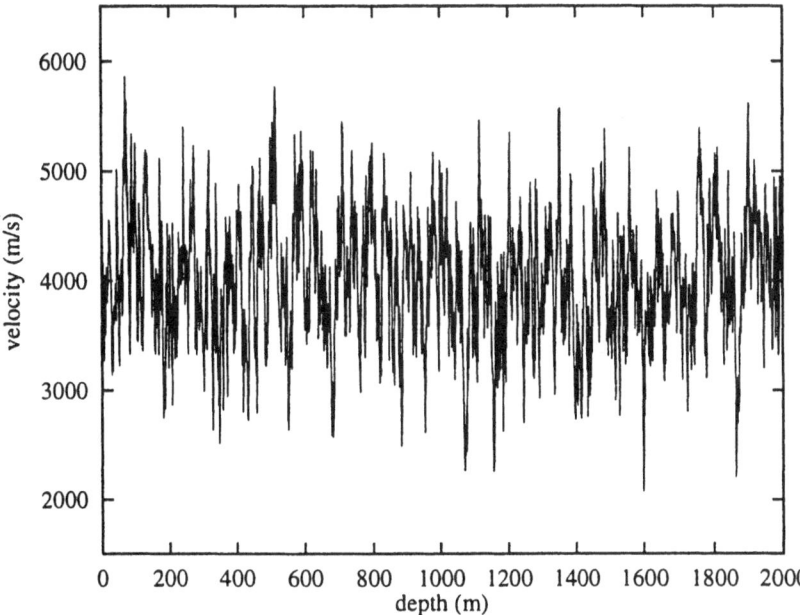

Fig. 2. Synthetic sonic log with an exponential correlation function. The averaged compressional velocity is $4000 m/s$, the correlation distance is 5m and the standard deviation is 15 percent. (From Shapiro and Hubral, 1996).

In the seismic practice L is still smaller than the localization length, which is of the order of 10 - 100 km or even larger for usual sediments. However, $L \gg l$, where l is a correlation length of the stratification. Under such conditions the wavefield has not yet been decreased very strongly. Its approximately exponential attenuation is due to a not yet strongly developed localization. This is so-called *weak localization*. The attenuation associated with it will be observed in each individual realization of the random medium. However, the corresponding attenuation coefficient γ has not yet reached the limit given by equation (2.16).

Attenuation in the context of strong and weak localization refers to the behavior of the energy of the transmitted wavefield. It must be strictly distinguished from the attenuation of the meanfield (i.e., the ensemble-averaged transmissivity). This is partially caused by the different phase shifts of the transmissivities for the different realizations of the random medium. Attenuation of the meanfield is a statistical effect, which exists even in cases where no attenuation due to localization can be observed. Consequently meanfield attenuation does not describe the attenuation coefficient γ in a single realization of a 1-D random medium.

Using the example of the pressure wavefield in a 1-D heterogeneous fluid given above we can now more precisely define the effect of the stratigraphic filtering which we are going to describe. Let us denote as $T(z)$ the down-going part of the pressure wavefield p_{true}. Clearly, at $z \leq 0$ the wavefield $T(z)$ coincides with the incident plane wave. At $z \geq L$ this is the transmitted plane wave. Thus, we call this wavefield the transmissivity. Then, in the case of the oblique incidence (the geometry is shown in Figure 1) we will try to approximate the transmissivity by the following formula:

$$
\begin{aligned}
T(\omega, p, x, z, t)|_{z=L} = T(\omega, p, x, z, t)|_{x,z,t=0} \times \\
\times \exp\{i(\psi(\omega, p)L + \omega p x - \omega t) - \gamma(\omega, p)L\},
\end{aligned}
\tag{2.17}
$$

where ω is the circular frequency and ωp is the horizontal wavenumber. Quantity γ is an approximation of the attenuation coefficient discussed above. Quantity ψ is the vertical phase increment discussed in the next section. Thus, the exponential factor

$$
F_{str}(\omega) \equiv \exp\{i\psi(\omega, p)L - \gamma(\omega, p)L\}
\tag{2.18}
$$

describes the linear-filter effect of the stratified medium, which is referred to as stratigraphic filtering. One of our main problems will be to find simple accurate expressions for the quantities ψ and γ.

2.5   Self-Averaged Quantities

A self-averaged quantity is a certain wavefield attribute, which assumes its mean value in one typical realization. A realization of a random medium is called typical if it belongs to a subset of realizations, which has the probability one. This implies that theoretically there may exist a subset of non-typical realizations of a random medium, for which a self-averaged quantity does not assume its mean value. The probability of this subset to exist, however, is zero.

For a self-averaged quantity there is no need for ensemble averaging provided the wave passes through a sufficiently thick typical realization of the medium. In this case the value of a self-averaged quantity is the same as its ensemble-averaged value. In other words the expected value of a self-averaged quantity, provided by the theory based on ensemble averaging, can be measured directly in one realization.

Lifshits et al. (1988) proved that the *attenuation coefficient* of the time-harmonic transmissivity, γ, is a self–averaged quantity. Formally, this means that the limiting value in equation (2.16) is not random. It is in fact identical for all typical realizations of the random medium and it is also equal to its mathematical expectation. The second self–averaged quantity turns out to be the (unwrapped) phase of the time-harmonic transmissivity $T(\omega, x = 0, z = L)$ divided by L. This is referred to as the *vertical phase increment*, which we denote by ψ:

$$\psi = \frac{\arg\left(T(\omega, x = 0, z = L)\right)}{L}.$$

The fact that γ and ψ are self-averaged quantities can be quickly appreciated with the help of the following computational modeling: Construct various typical realizations of a random medium of length L and compute both values. If L increases to infinity one will observe that γ will converge to the same fixed value, constant in all realizations. The same happens also with ψ.

Figure 3 provides an illustration of the self-averaging effect. The dotted wiggly curve shows the values of γ numerically computed for a model with $L = 200m$. The statistical characteristics of the random medium are given in section 6.3. The diamonds show the values of γ computed for the same model, however with $L = 2000m$. It is easy to observe that the fluctuations of γ are smaller for a larger L. Thus, even though γ has still not yet reached its limiting value in the sense of equation (2.16), it can be approximately considered as a self-averaged quantity. Note that the solid line is calculated by the generalized O'Doherty-Anstey formula (4.18).

It has been shown (Lifshits et al. 1988) that self-averaging is a rather general effect. It occurs in the case of large as well as small fluctuations of medium parameters. This effect controls the statistics of self-averaged quantities, which can be different properties of different physical systems with randomness (so-called *systems with disorder*). Typically the self-averaging is observed in an experiment where the size of a system (for wave propagation in 1-D random medium this is *the distance L*) is changed.

The physical reason for self-averaging can be explained as follows. The attenuation and phase shift of the plane-wave time-harmonic transmissivity essentially involve an *integration* of medium parameters. This integration (after normalizing by the medium size L) is equivalent to the spatial averaging, which yields the same results as the ensemble averaging for ergodic random media. Therefore, the quantities γ and ψ tend to their mathematical expectations, i.e. they are self-averaged quantities. The larger L the better is this spatial averaging. The deviations of the self-averaged quantities from their mathematical expectations converge to zero with $1/\sqrt{L}$. Formula (6.13) from this book for the relative standard deviation of γ describes exactly this behavior. In fact this formula can be

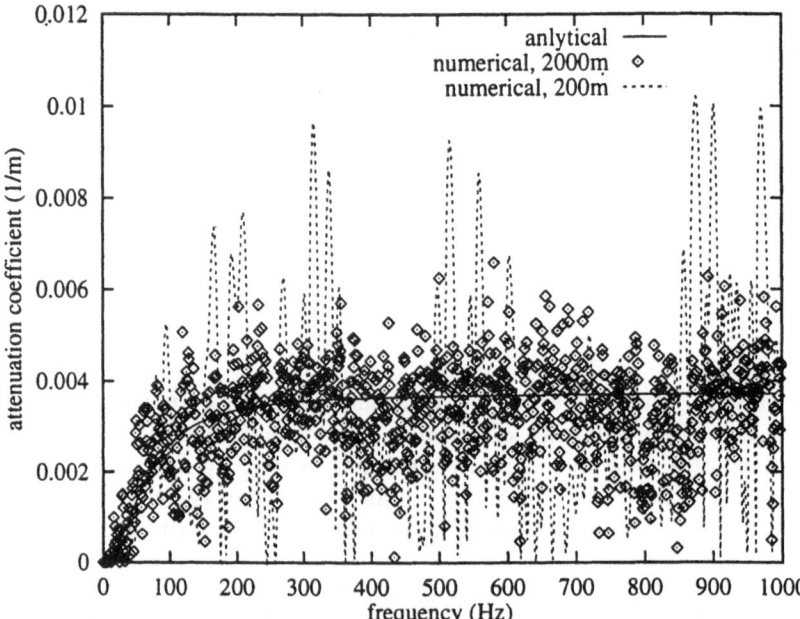

Fig. 3. Attenuation coefficient of the transmissivity of the vertically-incident P-wave. The dotted line shows results numerically computed in the case of 200 random layers. The diamonds show results of the same computations but now in the case of 2000 random layers (the thickness of each layer is equal to 1m). The smooth solid line is computed by the generalized O'Doherty-Anstey formula. The parameters of the medium are given in section 6.3.

considered as a proof (for the case of small fluctuations of the medium parameters only) that γ is a self-averaged quantity.

2.6   Products of Random Matrices

It should be pointed out that self-averaging and localization are physical effects. They are more general than the O'Doherty-Anstey formulas because they are not limited to small-fluctuation assumptions. They are neither limited to the low-frequency range. Mathematically they both are consequences of the asymptotic properties of products of random matrices.

It is known that in multilayered media the wavefield, say g, at two depths $z = 0$ and $z = \zeta$ is linked by the matrix product

$$R_N = P_N P_{N-1} \cdots P_2 P_1 \tag{2.19}$$

of so-called propagator matrices P_i (for examples of g and matrix P_i see equations (5.8) and (5.17), respectively):

$$g(\zeta) = R_N \, g(0). \tag{2.20}$$

The propagator matrices characterize the up- and downgoing wave propagation in each elementary homogeneous layer (see Section 5.3 and references there; see also, Aki, Richards 1981, pp. 273-286). Using them the elastic transmissivity for a random medium can be expressed in terms of products of random matrices. By multiplying the propagator matrices of subsequences of matrices for layered stacks, which are thicker than the correlation length, the product of random propagator matrices can be reduced to a product of *independent identically distributed matrices* (i.e., independent random matrices with equal probability-density functions).

Such a product R_N of N random independent identically distributed $D \times D$ matrices has asymptotic properties (for more details and references see the book of Crisanti et al. 1993, pp. 18 - 22) as described by the following theorems:

- *Fürstenberg's theorem.* This states that the limit

$$\gamma_1 = \lim_{N \to \infty} \frac{1}{N} \ln |R_N x| \tag{2.21}$$

exists and is not random for almost all D-dimensional vectors x. The quantity γ_1 is called the *Lyapunov characteristic exponent.*

- *Multiplicative ergodic theorem of Oseledets.* This states that the limiting matrix

$$V = \lim_{N \to \infty} \left(R_N{}^\dagger R_N \right)^{1/(2N)} \tag{2.22}$$

exists and has D positive non-random eigenvalues $\exp(\gamma_i)$ (here $\dagger$ stands for hermitian conjugate). The exponents $\gamma_1 \geq \gamma_2 \geq ... \geq \gamma_D$ are called *Lyapunov characteristic exponents.*

- *Virtser's theorem.* This states that for the product of a sequence of independent identically distributed $2D \times 2D$ symplectic random matrices

$$\gamma_1 > \gamma_2 > ... > \gamma_D > 0, \quad \gamma_{2D-i+1} = -\gamma_i. \tag{2.23}$$

One can notice that these theorems are similar to the well-known *central-limit theorem* for sums of random variables (see, e.g., Korn and Korn, 1961, Section

18.6-5.b). By direct calculations it can be also shown that for elastic waves the $2D \times 2D$ propagator matrix $\mathbf{P}$ is symplectic, i.e.,

$$\mathbf{P}^T \mathbf{J} \mathbf{P} = \mathbf{J} \equiv \begin{pmatrix} \mathbf{0} & \mathbf{I} \\ -\mathbf{I} & \mathbf{0} \end{pmatrix}, \tag{2.24}$$

where $\mathbf{0}$ and $\mathbf{I}$ denote $D \times D$ zero- and identity matrices, respectively. Moreover, the propagator matrices make up the so-called *symplectic group*, i.e., the product of any two propagator matrices is also symplectic. Therefore, the above theorems imply that plane seismic wavefields are localized in random media (from Virtser's theorem) and their attenuation coefficients are self-averaged quantities (from Oseledets's theorem).

Therefore, it is not surprising that different forms of self-averaging, e.g., the transient-transmissivity *pulse stabilization*, which will be demonstrated by numerical experiments (see Figures 27, 28, 29 and 30) can be found and strictly proven for particular cases such as the small-fluctuation or low-frequency approximation (Lewicki, 1994, and Lewicki et al. 1994).

If $D = 1$, there will exist only a single mode in the wavefield. A single-mode propagation is described later in Chapters 3 and 4 (see also introduction into Chapter 5). In this case the Fürstenberg's theorem states the existence of the localization length for the wavefield. The localization length is proportional to $1/\gamma_1$. Interestingly, this theorem describes not the attenuating but rather the exponentially growing solutions of wave equations. A simultaneous consideration of this theorem with the Virtser's theorem shows that there exist an exponentially decreasing solution with a Lyapunov characteristic exponent of exactly the same absolute value. It is precisely this solution, which describes the actually existing wavefield satisfying the physically meaningful boundary conditions (see also Section 4.2).

In the case $D > 1$ (multimode propagation; e.g., $D = 2$ for P-SV elastic waves described in Chapter 5) the Fürstenberg's theorem assures the existence of the largest Lyapunov characteristic exponent only. Realistic sources radiate usually mixed (i.e., multimode) wavefields. Such wavefields will be completely localized only if their mode with the smallest Lyapunov characteristic exponent, i.e., with the largest localization length, is localized. Therefore, in the case $D > 1$ the localization length of the complete wavefield is almost always (in the sense of the Fürstenberg's theorem) proportional to $1/\gamma_D$ (due to the Virtser's theorem).

In the multimode case one can imagine a situation, where some modes have been already localized, while others still propagate. Then, the relationships between the energies of localized modes and propagating modes will not be anymore controlled by the wavefield source. This relationship will be defined by the local conditions of the mode conversion in each point of the medium. For the case

$D = 2$ such a situation, called the *equilibration*, was analyzed in Kohler et al. (1996a; 1996b).

However, the generalized O'Doherty-Anstey approximation works in the spatial domain, where all modes are still propagating (see also Section 5.7). This is typical for the weak localization. Thus, the generalized O'Doherty-Anstey formulas provide small-fluctuation description of weakly localized wavefields.

3   Normal-Incidence Waves in a Stack of Layers

As we understood from the previous chapter, in multi-layered sediments the effects of multiple scattering are of importance. To analyze them we turn to the classical model of n homogeneous horizontal layers embedded between two homogeneous (not necessarily equivalent) half-spaces. We consider the propagation of a normally incident plane wave. Due to this our consideration will be valid for waves of any nature (e.g., pressure waves or elastic P- as well as S-waves), where 1-D propagation can be described as a single-mode *scalar wave*. The homogeneous layers constitute a so-called *Goupillaud model* (Goupillaud, 1961). In this model the thicknesses of the layers and the corresponding propagation velocities v_i and densities ϱ_i may differ. However, the normal-incidence travel times in all layers are the same.

This model corresponds to discretizing the depth-dependent compressional (or shear) velocity and the density of rocks in the time domain rather than in the depth domain. If the discretizing interval is sufficiently small (e.g., much smaller than the correlation length of the inhomogeneities), then both models (the model composed of layers of equal travel time and the model composed of layers of equal thickness) are equivalent for the wave propagation. The applicability of the Goupillaud model is, of course, restricted to one-dimensional wave propagation. However, this restriction is not too critical for sedimentary basins, where often the structure of inhomogeneities is approximately one-dimensional, at least in the first Fresnel volume.

In spite of the restrictions mentioned above, Goupillaud models are often and successfully used in seismics and seismology to simulate the reflection and the transmission of a plane wave normally incident onto a multi-layered geological structure (see e.g., Robinson and Treitel, 1980; Menke and Chen, 1984; Burridge et al., 1988; etc). Beside the practical importance of understanding plane-wave propagation in such a system for problems of *seismic deconvolution* or inversion, the Goupillaud model is very suitable to obtain a better physical picture of the multiple-scattering effects. It is also possible to develop with its help higher-order scattering approximations of the transmitted and the reflected wavefields. Such approximations are valid even in single realizations of random media. Using this simple model such important details of seismograms like *generalized primaries* and their *coda* as well as the nature of the O'Doherty-Anstey approximation can be better understood.

In this chapter we follow the paper of Shapiro and Treitel (1997), where the classical formalism of polynomial representations of $z-$transformed wavefields was combined with methods of statistical wave-propagation theory. Starting with the transmissivity we obtain an exact representation for it as well as the approximation (3.31), which is very good, and generally, more accurate than the O'Doherty-Anstey formula (3.32). This approximation describes the main part of the transmitted signal (generalized primaries) as well as an earlier part of its coda. For stochastic stratifications our analysis shows that, whereas *the generalized primaries are controlled by the second statistical moments* of the reflection coefficient series, *the coda is controlled by its fourth statistical moments*.

Note, that in this chapter and in Section 10.1 the variable z denotes the argument of the z-transforms of the reflectivity and transmissivity. In other parts of the book z denotes the depth. The reason for this is that in this chapter and in Section 10.1 we consider the wave propagation in a coordinate system related to the traveltime. In other parts of the book, wavefields are considered in coordinate systems related to the depth.

3.1   Recursive Description of the Wavefield

We consider now the transmissivity of a transient plane wave normally incident from above onto n homogeneous horizontal layers embedded between two homogeneous half-spaces (see Figure 1). In such a model the transmissivity T is the wave transmitted into the lower half-space and the reflectivity R is the wave reflected into the upper half-space. The homogeneous layers compose a Goupillaud model. We will numerate layers by the numbers $i = 1, 2, 3, ..., n$. The upper half-space, from which the δ-pulse plane wave impinges, has the number 0. The lower half-space has the number $n + 1$. In the medium there are $n + 1$ interfaces, which we will numerate by the numbers $i = 0, 1, 2, ..., n$. Therefore, a layer with the number i has a top interface with the number $i - 1$ and a bottom interface with the number i. The layers are characterized by the constant τ called the layer time and which is the two-way normal incidence travel time in each layer.

Following Robinson and Treitel (1977) we characterize each interface i by its downward reflection and transmission coefficients c_i and t_i, respectively. For example, in the case of an incident plane compressional wave from above these coefficients for the displacement are:

$$c_i = \frac{Z_i - Z_{i+1}}{Z_i + Z_{i+1}}, \qquad t_i = 2\frac{Z_i}{Z_i + Z_{i+1}}, \tag{3.1}$$

where $Z_i = v_i \varrho_i$ is the *acoustic impedance* in the layer of number i.

Using a formalism similar to the matrix-propagator approach, Robinson (1967, p. 125) has shown that in such a system the z-transforms of the δ-pulse transmissivity $T(z)$ and of the δ-pulse reflectivity $R(z)$ are

$$T(z) = \frac{t_0 t_1 ... t_n}{P_n(z) - c_0 Q_n(z)} z^{n/2}, \qquad R(z) = \frac{c_0 P_n(z) - Q_n(z)}{P_n(z) - c_0 Q_n(z)}, \tag{3.2}$$

where the quantities $P_n(z)$ and $Q_n(z)$ are the *fundamental polynomials*. These are given by the following recursive relationships:

$$P_n(z) = P_{n-1}(z) - c_n z^n Q_{n-1}(z^{-1}), \quad Q_n(z) = Q_{n-1}(z) - c_n z^n P_{n-1}(z^{-1}), \tag{3.3}$$

with $P_0 = 1$ and $Q_0 = 0$.

Formulas (3.2) provide the polynomial representations of the *z-transforms* of the transmitted and reflected wavefields

$$\begin{aligned} T(z) &= z^{n/2}(T_0 + T_1 z + T_2 z^2 + ...), \\ R(z) &= R_0 + R_1 z + R_2 z^2 + ..., \end{aligned} \tag{3.4}$$

where the time series $T_0, T_1, T_2, ...$ and $R_0, R_1, R_2, ...$ represent the seismograms, which are observed in the lower-, respectively, upper half-spaces. The time increment between samples of both seismograms is τ. The substitution $z = e^{-i\omega\tau}$ turns equations (3.4) into the discrete Fourier transforms of the transmissivity and reflectivity.

It is important to note that formulas (3.2) are exact and, therefore, they include all effects of multiple scattering. However, they express the reflectivity and transmissivity in terms of recursively computed polynomials. Using these formulas in the next sections, we derive exact explicit formulas of the transmissivity and reflectivity.

3.2   Exact Expressions for the Wavefield

For the sake of convenience we consider, instead of the fundamental polynomials P_n and Q_n the following two polynomials:

$$
\begin{aligned}
A_n(z) &= P_n(z) - c_0 Q_n(z), \\
B_n(z) &= c_0 P_n(z) - Q_n(z).
\end{aligned}
\tag{3.5}
$$

In terms of these polynomials the z-transforms of the transmissivity and the reflectivity read:

$$
T(z) = \frac{t_0 t_1 \ldots t_n}{A_n(z)} z^{n/2}, \qquad R(z) = \frac{B_n(z)}{A_n(z)}.
\tag{3.6}
$$

The fundamental polynomials have then the following forms:

$$
P_n(z) = \frac{A_n(z) - c_0 B_n(z)}{1 - c_0^2}, \qquad Q_n(z) = \frac{c_0 A_n(z) - B_n(z)}{1 - c_0^2}.
\tag{3.7}
$$

Using these equations along with the recursion (3.3) and definitions (3.5) we obtain

$$
\begin{aligned}
A_n(z) &= A_{n-1}(z) + c_n z^n B_{n-1}(z^{-1}), \\
B_n(z) &= B_{n-1}(z) + c_n z^n A_{n-1}(z^{-1}).
\end{aligned}
\tag{3.8}
$$

This recursive system can be reduced to the following two independent recursive equations:

$$
A_n(z) = A_{n-1}(z) + z c_n c_{n-1}^{-1}(A_{n-1}(z) - A_{n-2}(z)) + z c_n c_{n-1} A_{n-2}(z).
\tag{3.9}
$$

$$
B_n(z) = B_{n-1}(z) + z c_n c_{n-1}^{-1}(B_{n-1}(z) - B_{n-2}(z)) + z c_n c_{n-1} B_{n-2}(z).
\tag{3.10}
$$

Formulas (3.9) and (3.10) along with equalities $A_0 = 1$, $B_0 = c_0$, $A_1 = 1 + c_0 c_1 z$ and $B_1 = c_0 + c_1 z$ provide an efficient way of numerical computations of the reflectivity and the transmissivity when using relationships (3.6). It is clear that all effects of multiple scattering are described fully by this recursive formalism. Later in this chapter we use these formulas in some of our analytical considerations as well as in all numerical examples.

Let us now introduce two new polynomials:

$$A_n^R(z) = z^n A_n(z^{-1}), \qquad B_n^R(z) = z^n B_n(z^{-1}). \tag{3.11}$$

The recursive equations for them can be obtained from equations (3.8):

$$A_n^R(z) = z A_{n-1}^R(z) + c_n B_{n-1}(z),$$
$$B_n^R(z) = z B_{n-1}^R(z) + c_n A_{n-1}(z). \tag{3.12}$$

Both systems of equations, (3.8) and (3.12) can be combined into the following matrix recursion:

$$\begin{pmatrix} A_n & B_n^R \\ B_n & A_n^R \end{pmatrix} = \begin{pmatrix} A_{n-1} & B_{n-1}^R \\ B_{n-1} & A_{n-1}^R \end{pmatrix} \begin{pmatrix} 1 & c_n \\ c_n z & z \end{pmatrix}. \tag{3.13}$$

Therefore, we obtain:

$$\begin{pmatrix} A_n & B_n^R \\ B_n & A_n^R \end{pmatrix} = \begin{pmatrix} 1 & c_0 \\ c_0 & 1 \end{pmatrix} \begin{pmatrix} 1 & c_1 \\ c_1 z & z \end{pmatrix} \begin{pmatrix} 1 & c_2 \\ c_2 z & z \end{pmatrix} \cdots \cdots \begin{pmatrix} 1 & c_n \\ c_n z & z \end{pmatrix}. \tag{3.14}$$

The determinant of the left- and right-hand sides of this equation provides the expression of energy conservation:

$$|A_n|^2 - |B_n|^2 = (1 - c_0^2)(1 - c_1^2)...(1 - c_n^2). \tag{3.15}$$

Multiplying now the first equation in (3.8) with $B_{n-1}(z)$ and the second one with $A_{n-1}(z)$, extracting from the second equation the first one and taking into account property (3.15), we obtain:

$$B_n(z)A_{n-1}(z) - B_{n-1}(z)A_n(z) = z^n c_n (1 - c_0^2)(1 - c_1^2)...(1 - c_{n-1}^2). \tag{3.16}$$

This can be rewritten in the more instructive form

$$\frac{B_n(z)}{A_n(z)} = \frac{B_{n-1}(z)}{A_{n-1}(z)} + \frac{z^n c_n (1 - c_0^2)(1 - c_1^2)...(1 - c_{n-1}^2)}{A_n(z) A_{n-1}(z)}, \tag{3.17}$$

which is a recursive formula for the reflectivity. Taking into account the initial conditions, i.e., $A_0 = A_0^R = 1$ and $B_0 = B_0^R = c_0$, we can solve this recursion:

$$R(z) = c_0 + \sum_{j=1}^{n} \frac{z^j c_j (1 - c_0^2)(1 - c_1^2)...(1 - c_{j-1}^2)}{A_j(z) A_{j-1}(z)}. \tag{3.18}$$

This interesting formula was first derived by Hubral, Treitel and Gutowski in a different way (Hubral et. al, 1980). It clearly demonstrates the key role of the polynomial A_n, which controls the transmissivity, which in turn appears in the description of the reflectivity. What we have in each term of the sum (3.18) is a product of two z-transformed transmissivities.

Therefore, the particular recursion for the polynomial A_n is important. From equations (3.8), (3.12) we obtain:

$$A_n(z) = 1 + z \sum_{j=1}^{n} c_j B_{j-1}^R(z),$$
$$B_n^R(z) = z^n + \sum_{j=1}^{n} c_j z^{n-j} A_{j-1}(z). \tag{3.19}$$

Substituting the expression for B_n^R into that for A_n, we obtain a closed-form equation for the polynomial A_n:

$$A_n(z) = 1 + \sum_{j=1}^{n} \sum_{m=0}^{j-1} c_j c_m z^{j-m} A_{m-1}(z), \tag{3.20}$$

where we defined $A_{-1}(z) = 1$.

Using equation (3.20) we can show by mathematical induction that the following formula for the polynomial $A_n(z)$ is exactly valid:

$$A_n = 1$$

$$+ \sum_{i_1=1}^{n} c_{i_1} \sum_{i_2=0}^{i_1-1} c_{i_2} z^{i_1-i_2}$$

$$+ \sum_{i_1=3}^{n} c_{i_1} \sum_{i_2=2}^{i_1-1} c_{i_2} z^{i_1-i_2} \sum_{i_3=1}^{i_2-1} c_{i_3} \sum_{i_4=0}^{i_3-1} c_{i_4} z^{i_3-i_4}$$

$$+ \ \ldots\ldots\ldots\ldots\ldots\ldots \tag{3.21}$$

$$+ \sum_{i_1=2N-1}^{n} c_{i_1} \sum_{i_2=2N-2}^{i_1-1} c_{i_2} z^{i_1-i_2} \sum_{i_3=2N-3}^{i_2-1} c_{i_3} \sum_{i_4=2N-4}^{i_3-1} c_{i_4} z^{i_3-i_4} \ldots$$

$$\ldots \sum_{i_{2N-1}=1}^{i_{2N-2}-1} c_{i_{2N-1}} \sum_{i_{2N}=0}^{i_{2N-1}-1} c_{i_{2N}} z^{i_{2N-1}-i_{2N}} \ ,$$

where $N = Int((n+1)/2)$ and $Int(x)$ denotes the integer part of x.

By substituting now the polynomial A_n from formula (3.21) into the first equation of system (3.6) and into equation (3.18), we obtain exact explicit formulas for the z−transforms of the reflectivity and the transmissivity. It is clear, however, that further simplifications are desirable to better understand the structure of seismograms. In the following sections we introduce some useful approximations taking a certain important part of multiple-scattering effects into account, thus simplifying the expressions for the transmissivity and the reflectivity.

3.3   Explicit Approximations for the Transmissivity

We see now that the polynomial $A_n(z)$ is a combination of terms of the form

$$\sum_{j=a}^{b} c_j \sum_{l=a-1}^{j-1} c_l z^{j-l}. \tag{3.22}$$

By the change of the summation indices from j and l to $k = j - l$ and l, such terms are transformed to the form

$$\sum_{k=1}^{b-a+1} \sum_{l=a-1}^{b-k} c_l c_{l+k} z^k. \tag{3.23}$$

Using this transformation we obtain another, still exact, expression for the polynomial $A_n(z)$:

$$
\begin{aligned}
A_n = 1 & \\
+ & \sum_{k_1=1}^{n} \sum_{i_1=0}^{n-k_1} c_{i_1} c_{i_1+k_1} z^{k_1} \\
+ & \sum_{k_1=1}^{n-2} \sum_{i_1=2}^{n-k_1} c_{i_1} c_{i_1+k_1} z^{k_1} \sum_{k_2=1}^{i_1-1} \sum_{i_2=0}^{i_1-k_2} c_{i_2} c_{i_2+k_2} z^{k_2} \\
+ & \ \dots\dots\dots\dots \\
+ & \sum_{k_1=1}^{n-2N+2} \sum_{i_1=2N-2}^{n-k_1} c_{i_1} c_{i_1+k_1} z^{k_1} \sum_{k_2=1}^{i_1-2N+3} \sum_{i_2=2N-4}^{i_1-1-k_2} c_{i_2} c_{i_2+k_2} z^{k_2} \dots \\
& \dots \sum_{k_N=1}^{i_{N-1}-1} \sum_{i_N=0}^{i_{N-1}-1-k_N} c_{i_N} c_{i_N+k_N} z^{k_N}.
\end{aligned}
\tag{3.24}
$$

In the following analysis we compare this formula with the Taylor expansion of the exponential function:

$$
\begin{aligned}
E_n = \exp\Big(& \sum_{k=1}^{n} \sum_{i=0}^{n-k} c_i c_{i+k} z^k \Big) = 1 \\
+ & \sum_{k_1=1}^{n} \sum_{i_1=0}^{n-k_1} c_{i_1} c_{i_1+k_1} z^{k_1} \\
+ & \frac{1}{2!} \sum_{k_1=1}^{n} \sum_{i_1=0}^{n-k_1} c_{i_1} c_{i_1+k_1} z^{k_1} \sum_{k_2=1}^{n} \sum_{i_2=0}^{n-k_2} c_{i_2} c_{i_2+k_2} z^{k_2} \\
+ & \ \dots\dots\dots\dots \\
+ & \frac{1}{N!} \sum_{k_1=1}^{n} \sum_{i_1=0}^{n-k_1} c_{i_1} c_{i_1+k_1} z^{k_1} \sum_{k_2=1}^{n} \sum_{i_2=0}^{n-k_2} c_{i_2} c_{i_2+k_2} z^{k_2} \dots \\
& \dots \sum_{k_N=1}^{n} \sum_{i_N=0}^{n-k_N} c_{i_N} c_{i_N+k_N} z^{k_N} \\
+ & \ \dots\dots\dots\dots \ .
\end{aligned}
\tag{3.25}
$$

In spite of a clear difference between both series (e.g., the first one is finite and the second one is infinite) there is also a profound similarity between them. Obviously, each term of the polynomial (3.24) describes certain sequences of elementary reflections of waves from the corresponding interfaces. The first two terms of equation (3.24) just coincide with the first two terms of the exponential series. These two terms can be interpreted as a single-scattering approximation of the polynomial A_n. Its validity requires that

$$n^2\varepsilon^2 \ll 1, \tag{3.26}$$

where $c_i = O(\varepsilon)$. In this range polynomials (3.24) and (3.25) are approximately equal.

One can, however, see that not only the two first terms of polynomials (3.24) and (3.25) coincide. There is as well a lot of similarity between their other terms (in the following we call them higher-order terms). This observation is important because the higher-order terms of the polynomial A_n describe multiple scattering effects. In order to prove the similarity of the higher-order terms one could directly compare the corresponding lines of equations (3.24) and (3.25). In the appendix to this chapter, however, we do this in a different way. We show that

$$A_n \approx E_n, \tag{3.27}$$

if

$$n\varepsilon^2 \ll 1. \tag{3.28}$$

This validity range is much broader than range (3.26). Therefore, approximation (3.27) includes higher-order multiple-scattering effects.

Using approximation (3.27) of the polynomial $A_n(z)$ we obtain the following approximation of the transmissivity (with the same precision, as later specified by condition (3.63)):

$$T(z) \approx t_0 t_1 ... t_n z^{n/2} \exp(-\sum_{k=1}^{n}\sum_{i=0}^{n-k} c_i c_{i+k} z^k). \tag{3.29}$$

This approximation, however, has been already based on inequality (3.58). Therefore, the following Taylor expansion will not reduce the precision:

$$\ln t_i = \ln(1 + c_i) \approx c_i - c_i^2/2. \tag{3.30}$$

Taking this into account we obtain:

$$T(z) \approx z^{n/2} \exp\left[\sum_{i=0}^{n}(c_i - c_i^2/2) - \sum_{k=1}^{n}\sum_{i=0}^{n-k} c_i c_{i+k} z^k\right]. \tag{3.31}$$

By assuming that the average of the reflection coefficients is equal to zero (this assumption implies that $\sum_{i=0}^{n} c_i = 0$ for a given n) approximation (3.29) is immediately reduced to the *deterministic form of the classical O'Doherty-Anstey formula* (for short: ODA formula):

$$T(z) \approx z^{n/2} \exp\left[-\sum_{i=0}^{n} c_i^2/2 - \sum_{k=1}^{n} \sum_{i=0}^{n-k} c_i c_{i+k} z^k \right]. \tag{3.32}$$

While O'Doherty and Anstey (1971) had to assume a statistical character of reflection coefficients, we have here just shown that this assumption is not needed. In fact, as we can see, the ODA formula results from (3.29) on the basis of a completely deterministic derivation of the transmissivity for a deterministic reflection-coefficient series.

Approximations (3.29) and (3.32) allow us to describe the earlier portions of the codas following the first arrival onset. For a reflection-coefficient series with zero average, if $n\varepsilon^2 = const$, such earlier portions of the transmissivity will tend to the exact series $T_0, T_1, \ldots$ in the limit $n \to \infty$. This convergence condition has been found and discussed by Burridge et al. (1988) and Berlyand and Burridge (1995). Nevertheless, for finite n, not too small ε and non-zero-average reflection coefficients, the results (3.29) and (3.31) are more accurate than the ODA formula (3.32) (see Figure 4).

3.4   Generalized Primary and Coda

From the above considerations we see that the transmissivity is very different from multiplying the incident wavefield (above taken as a delta spike) simply by the product of all elementary transmission coefficients. The absolute value of the peak amplitude of the transmissivity is much larger than this product. Thus, the transmissivity is a result of interference of the directly transmitted first-arrival δ-pulse wave (which becomes negligibly small with increasing travel-distance) and a complex package of internally scattered waves. Therefore, it is in fact the multiply-scattered complex-package wavefield, which makes seismic reflections from deep reflectors (which can be viewed on account of formula (3.18) as transient two-way transmissivities from separated reflectors in the medium) visible for us.

It is intuitively clear that the main parts of seismic reflection events following closely the events' first arrivals are especially important for subsurface imaging

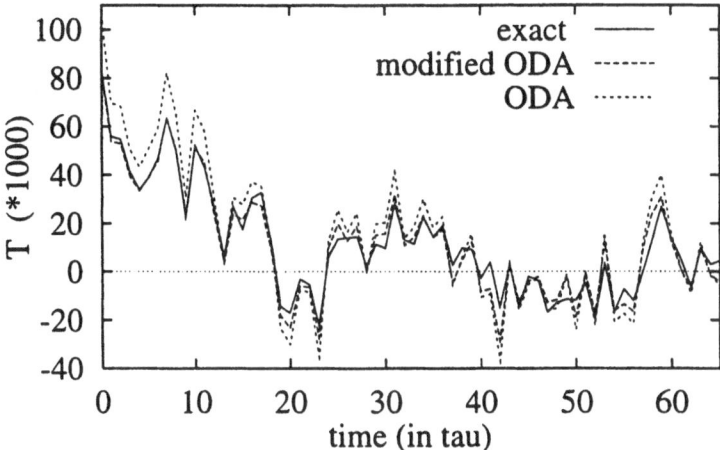

Fig. 4. Transmissivity for a stationary series of 500 reflection coefficients with $-0.45 < c_i < 0.45$ and the RMS value equal to 0.1. This series was obtained from a realization of an exponentially-correlated synthetic sonic log. The time delay is given relative to the first arrival and is measured in τ. The curves "modified ODA" and "ODA" were computed using equations (3.31) and (3.32), respectively. (Reprinted from Phys. of the Earth and Planet. Inter., v.104., p.152, 1997, Shapiro and Treitel, with kind permission from Elsevier Science - NL)

as they have the largest amplitudes. These parts of the seismic events we call *generalized primaries*. The later arriving rest we call the coda. Usually the codas are not very important for seismic reflection imaging. Moreover, sometimes they are even not visible.

Below we provide a more mathematically substantiated definition of generalized primaries and codas. Note that our definition here for the generalized primary is different from that given in Hubral et al (1980), where the generalized primaries is understood as the total two-way transient transmissivity from a reflector as given by each term in equation (3.18).

A convenient way to define the generalized primaries is a statistical one. In contrast to 2-D and 3-D randomly inhomogeneous media, application of the statistical approach to the wavefield description in single realizations of 1-D inhomogeneous media has some principal restrictions (see Chapter 2). Ensemble-averaged quantities, like e.g., the averaged field or its intensity may be not applicable to describing the features of single seismograms, like e.g., the coda of the transient transmissivity. In the following chapters we will show that averaged logarithms of the transmitted wavefield, obtained in the second Rytov approximation, can be used to obtain smooth versions of generalized-primary arrivals in single realizations of seismograms. Now, however, we would like to combine the polynomial-based deterministic formalism given above with the statistical **approach**.

Let us consider a reflection-coefficient series as a part of a single realization of an ergodic random process. We will use such a series as a model of reflection coefficients obtained from seismo-acoustic measurements in a borehole. Figure 5 shows reflection coefficients computed from a real-log data set. This data set was collected in a marine, hydrocarbon-productive sedimentary basin characterized by limestones, carbonates and coal/shale sequences.

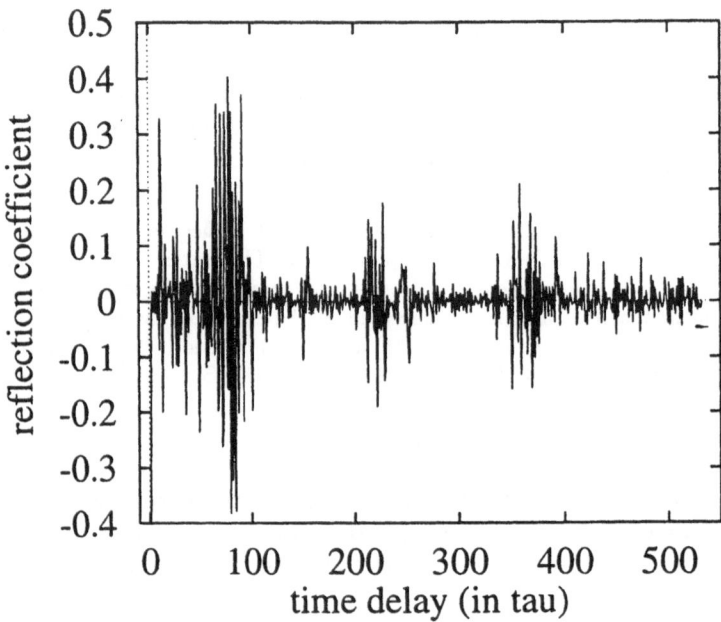

Fig. 5. Series of 531 P-wave reflection coefficients obtained from logs in a hydrocarbon-productive sedimentary basin. The first reflection coefficient (-0.4), corresponding to the ocean-bottom reflection, has been artificially added to the series. (Reprinted from Phys. of the Earth and Planet. Inter., v.104., p.153, 1997, Shapiro and Treitel, with kind permission from Elsevier Science - NL)

Usually, realistic reflection coefficients have a finite correlation. This means that the autocorrelation function

$$\varrho_c(l) = \lim_{n \to \infty} \left[\frac{1}{n} \sum_{i=0}^{n-1} c_i c_{i+l} \right] \tag{3.33}$$

significantly differs from zero for finite $l = O(a)$, where a is the correlation length of the reflection-coefficient series. Because of the assumed ergodicity the quantity $\varrho_c(k)$ is equal to the ensemble-averaged autocorrelation function. However, in all realistic situations the number n of the reflection coefficients is finite, and we have

$$\sum_{i=0}^{n-l} c_i c_{i+l} = (n - l + 1)\varrho_c(l) + O(\sqrt{Var[\sum_{i=0}^{n-l} c_i c_{i+l}]}), \tag{3.34}$$

where $Var[x]$ is the variance of the quantity x. This can be expressed in terms of statistical averages as follows:

$$\begin{aligned} Var\left[\sum_{i=0}^{n-l} c_i c_{i+l}\right] &=< \left[\sum_{i=0}^{n-l} c_i c_{i+l}\right]^2 > -(n - l + 1)^2 \varrho_c(l)^2 \\ &= \sum_{i=0}^{n-l}\sum_{k=0}^{n-l} < c_i c_{i+l} c_k c_{k+l} > -(n - l + 1)^2 \varrho_c^2(l). \end{aligned} \tag{3.35}$$

The average under the double sum is a *fourth statistical moment of the reflection-coefficient series*.

To estimate the order of this quantity let us assume that the reflection-coefficient series is a realization of a stationary Gaussian random process with zero average. Then equation (3.35) can be further simplified:

$$\begin{aligned} Var\left[\sum_{i=0}^{n-l} c_i c_{i+l}\right] &= \sum_{i=0}^{n-l}\sum_{k=0}^{n-l} [\varrho_c^2(i - k) + \varrho_c(i + l - k)\varrho_c(k + l - i)] \\ &= \sum_{k=0}^{n-l}\sum_{g=-k}^{n-l-k} [\varrho_c^2(g) + \varrho_c(l + g)\varrho_c(l - g)]. \end{aligned} \tag{3.36}$$

In the case of a small correlation length $(a \ll n)$ of the reflection-coefficient series we have:

$$Var[\sum_{i=0}^{n-l} c_i c_{i+l}] = O((n - l)\varepsilon^4). \tag{3.37}$$

Therefore, with increasing n the first term on the right-hand side of equation (3.34) tends to infinity as $O(n\varepsilon^2)$, while the second term increases only as $O(n^{1/2}\varepsilon^2)$. Thus, the sum on the left-hand side of equation (3.34) tends to its ensemble-averaged value (in the sense of the relative error). As a consequence we can approximate this sum in equation (3.27) by its ensemble-average. The following is then valid (without reducing the precision of equation (3.63); see, however, the following discussion of the case of finite n):

$$A_n \approx \exp nS_1, \tag{3.38}$$

where we introduced

$$S_1 = \sum_{l=1}^{\infty} \varrho_c(l)z^l. \tag{3.39}$$

The statistical approximation (3.38) can be used now to obtain different approximations of the transmissivity, corresponding to equations (3.29)-(3.32). For a given finite n these will be valid, however, only for that part of the transmissivity, which is close to the first arrival. In fact, the number l controls the power of z in the first series from equation (3.2) representing the z−transformed transmissivity, and, therefore, it controls the time delay of the corresponding samples of the seismograms. In the case of a given finite n, due to the finite correlation of the series c_i, only for small enough l the first term in (3.34) is larger than the second one. Formally this means that besides the restriction (3.28) for the statistical approximation to be applicable, the following inequality is to be satisfied:

$$|\varrho_c(l)| \gg \sqrt{Var[\sum_{i=0}^{n-l} c_i c_{i+l}]/(n-l)}. \tag{3.40}$$

This inequality can be used as a formal definition of the *length of the generalized primary* l_{gp}. It is *the maximal l for which inequality (3.40) is still valid. Thus, the part of the transient transmissivity with $l \leq l_{gp}$ is called the generalized primary. The rest of the transient transmissivity (with $l \geq l_{gp}$) is called the coda.* Note, that in many cases a good approximation of the right-hand part of inequality (3.40) is $\sqrt{Var[c_i c_{i+l}]/n}$.

To estimate the coda in a classical way (see e.g., Sato, 1995), we should compute from equation (3.29) an expectation of the envelope of the transmitted wavefield. However, here, to obtain a rough estimation of the coda in single realizations of seismograms, we can apply a more straight-forward approach.

Indeed, let us return again to equation (3.34). If the number l is large enough (in the sense of inequality (3.40)) the sum on the left-hand side of equation (3.34) will be equal to its fluctuations. It is precisely these fluctuations that produce the earlier part of the coda of the transient transmissivity. The order of these fluctuations can be estimated as $\sqrt{Var[\sum_{i=0}^{n-l} c_i c_{i+l}]}$. Using approximation (3.29) we can estimate the effect of these fluctuations in the time domain. To do this, we expand the exponential part of this formula into the Taylor series:

$$T(z) \approx t_0 t_1 ... t_n z^{n/2} (1 - \sum_{k=1}^{n} \sum_{i=0}^{n-k} c_i c_{i+k} z^k) \qquad (3.41)$$

The inverse $z-$transform provides the transmissivity in the time domain:

$$T(l) \approx -t_0 t_1 ... t_n \sum_{i=0}^{n-l} c_i c_{i+l}, \qquad (3.42)$$

where $T(l)$ is the $l-$th (after the first arrival) sample of the transient transmissivity. In the coda-time range the transmissivity fluctuates due to the fluctuations of the sum on the right-hand side of this relationship. Therefore, the following estimation can be used:

$$T_{coda}(l) \sim t_0 t_1 ... t_n \sqrt{Var[\sum_{i=0}^{n-l} c_i c_{i+l}]}. \qquad (3.43)$$

In contrast to approximations (3.29)-(3.32) this one is valid in a more restricted domain of n and ε than those given by inequality (3.28). Due to the Taylor expansion, which we applied to obtain relationship (3.42), its validity domain is restricted by the following two inequalities:

$$n^{3/2} \varepsilon^2 \ll 1, \qquad (3.44)$$

and

$$l < n. \qquad (3.45)$$

The last inequality implies that estimation (3.43) describes the earlier coda only. Taking into account equation (3.35) we conclude that in contrast to the generalized primaries, which are controlled by the second statistical moment (e.g., by the quantity S_1), the coda is controlled by the fourth statistical moments of the reflection-coefficient series. Finally, in the case of a Gaussian zero-average reflection-coefficient series the following approximation can be used:

$$T_{coda}(l) \sim \varrho_c(0)\sqrt{n-l} e^{-\varrho_c(0)n/2}. \qquad (3.46)$$

We should note here that from numerical simulations Menke and Chen (1984) derived an empirical rule for the coda decay. As in our consideration they used Goupillaud models for their computations. In contrast to our study, however, they considered the envelope of the later coda rather than the earlier coda (see restriction (3.44)).

In order to demonstrate the above approximations let us consider a series of reflection coefficients of a stationary medium with weakly fluctuating acoustic impedances $Z_i = Z(1 + \zeta_i)$, where ζ_i is a small quantity $\zeta_i \ll 1$ with zero statistical average $< \zeta_i >= 0$. If the quantity ζ_i has the autocorrelation function

$$< \zeta_i \zeta_{i+k} >= \varrho_\zeta(k), \tag{3.47}$$

then the autocorrelation function of the reflection-coefficient series will be

$$\varrho_c(k) =< c_i c_{i+k} >\approx \left[2\varrho_\zeta(k) - \varrho_\zeta(k-1) - \varrho_\zeta(k+1)\right]/4. \tag{3.48}$$

In the particular case of a medium with an exponentially-correlated weakly-fluctuating acoustic impedance we have:

$$\varrho_\zeta = \sigma_\zeta^2 \exp\left[-k/a\right]. \tag{3.49}$$

Then the autocorrelation function of the reflection coefficients is

$$\varrho_c(0) = \sigma_\zeta^2(1 - e^{-1/a})/2, \tag{3.50}$$

and

$$\varrho_c(k) = -\varrho_c(0)(1 - e^{-1/a})e^{(1-k)/a}/2 \tag{3.51}$$

for $k > 0$.

Using these expressions we obtain the quantity S_1 from equation (3.39). This provides the following statistical counterparts of approximations (3.29) and (3.32), respectively:

$$T(z) = t_0 t_1 ... t_n z^{n/2} \exp\left[n\varrho_c(1)\frac{z}{z\exp(-1/a) - 1}\right], \tag{3.52}$$

and

$$T(z) = z^{n/2} \exp\left[-n\varrho_c(0)/2 + \frac{n\varrho_c(1)z}{z\exp(-1/a) - 1}\right]. \qquad (3.53)$$

Figures 6 and 7 show the exactly computed transmissivity (formulas (3.6) and (3.9)), in comparison with results of relationships (3.52) and (3.46) for the actual reflection-coefficient series shown in Figure 5. For these computations we assumed that the series shown in Figure 5 is stationary and has the autocorrelation given by equations (3.50) and (3.51). The corresponding correlation length a was found from the values $\varrho_c(0)$ and $\varrho_c(1)$. The following estimations were obtained: $\sqrt{\varrho_c(0)} = 0.08$, $a = 1$, and $\sigma = 0.14$. Therefore, we found a very rough approximation of the real stratification by a medium with an exponentially-correlated acoustic impedance. In spite of this simplification a good agreement between the exact transmissivity and the rough statistic estimations can be observed.

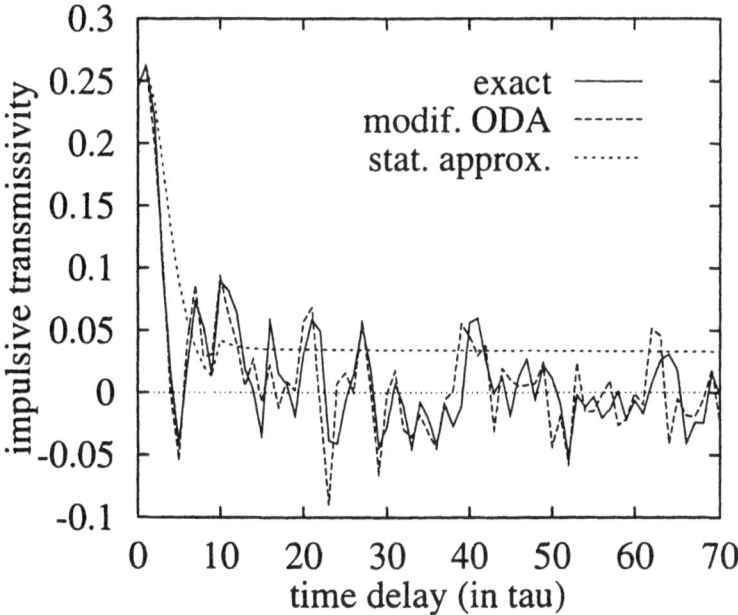

Fig. 6. The transient transmissivity for the reflection-coefficient series shown in Figure 5. The three lines shown are the exact transmissivity, the result of equation (3.29) and the combined result of equations (3.52) and (3.46), respectively. (Reprinted from Phys. of the Earth and Planet. Inter., v.104., p.155, 1997, Shapiro and Treitel, with kind permission from Elsevier Science - NL.)

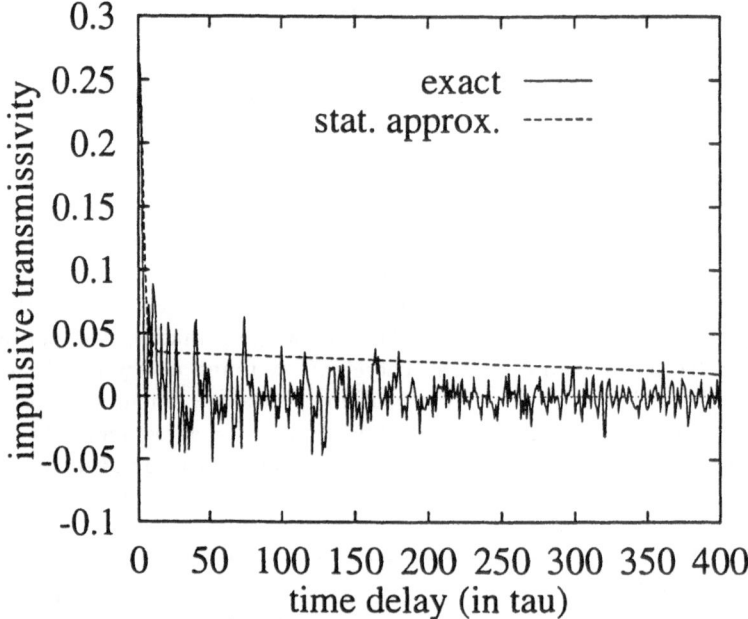

Fig. 7. The same as in Figure 6, however, on another time scale. The two curves show the exact transmissivity and the combined results of equations (3.52) and (3.46), respectively. (Reprinted from Phys. of the Earth and Planet. Inter., v.104., p.155, 1997, Shapiro and Treitel, with kind permission from Elsevier Science - NL).

3.5   Appendix: Approximation of the Polynomial A_n

In this appendix we show that polynomials (3.24) and (3.25) are approximately equal in a much broader validity range than indicated by the inequality (3.26). For this purpose we introduce the following polynomial:

$$P'_n = \sum_{k=1}^{n} \sum_{i=0}^{n-k} c_i c_{i+k} z^k . \tag{3.54}$$

By direct calculations one can show that

$$P'_n = P'_{n-1} + \sum_{k=1}^{n} c_{n-k} c_n z^k . \tag{3.55}$$

Using this formula to express P'_{n-1} in terms of P'_{n-2} and combining the resulting recursive equation for P'_{n-1} with the above one we obtain:

$$P'_n = P'_{n-1} + z(P'_{n-1} - P'_{n-2})c_n/c_{n-1} + zc_{n-1}c_n. \tag{3.56}$$

Substituting into this the equality $P'_n = \ln E_n$, we immediately obtain the recursive formula for the infinite polynomial E_n:

$$E_n = E_{n-1}(1 + \frac{E_{n-1} - E_{n-2}}{E_{n-2}})^{zc_n/c_{n-1}} \exp\left(zc_{n-1}c_n\right). \tag{3.57}$$

Now we make the two important assumptions:

$$\sigma \equiv \varepsilon^2 \ll 1, \tag{3.58}$$

and

$$\eta \equiv |\frac{E_{n-1} - E_{n-2}}{E_{n-2}}| \ll 1. \tag{3.59}$$

Later, we consider these assumptions in more details. With them we can replace the exponential and potential terms of equation (3.57) by the first two terms of their Taylor expansions:

$$E_n = E_{n-1}(1 + z\frac{c_n}{c_{n-1}}\frac{E_{n-1} - E_{n-2}}{E_{n-2}} + O(\eta^2))(1 + zc_{n-1}c_n + O(\sigma^2)). \tag{3.60}$$

This can be further transformed into

$$E_n = E_{n-1}(1 + z\frac{c_n}{c_{n-1}}\frac{E_{n-1} - E_{n-2}}{E_{n-1}} + zc_{n-1}c_n\frac{E_{n-2}}{E_{n-1}}) \times$$
$$\times (1 + O(\eta\sigma) + O(\eta^2) + O(\sigma^2)). \tag{3.61}$$

Finally we obtain

$$E_n = (E_{n-1} + zc_nc_{n-1}^{-1}(E_{n-1} - E_{n-2}) + zc_{n-1}c_nE_{n-2}) \times$$
$$\times (1 + O(\eta\sigma) + O(\eta^2) + O(\sigma^2)). \tag{3.62}$$

Taking into account that $E_0 = 1$ and $E_1 = 1 + c_0c_1z + O(\sigma^2)$ and comparing equation (3.62) with equation (3.9) we conclude that

$$A_n = E_n(1 + O(\eta\sigma) + O(\eta^2) + O(\sigma^2)). \tag{3.63}$$

Now we must consider restrictions (3.58) and (3.59). From inequality (3.59), using equation (3.55), we obtain

$$|\exp\sum_{k=1}^{n} c_{n-k}c_n z^k - 1| \ll 1. \tag{3.64}$$

Using the Taylor expansion of an exponential function this inequality can be reduced to

$$|\sum_{k=1}^{n} c_{n-k}c_n z^k| \ll 1, \tag{3.65}$$

which in turn can be roughly expressed as

$$n\varepsilon^2 \ll 1. \tag{3.66}$$

This inequality, of course, is stronger than inequality (3.58).

Under restriction (3.66) all terms of orders $O(\eta\sigma)$, $O(\eta^2)$, and $O(\sigma^2)$ are much smaller than 1 and the approximation

$$A_n \approx E_n \qquad\qquad\qquad (3.67)$$

is valid. However, a comparison between inequalities (3.26) and (3.66) immediately shows that the first one is much stronger than the second one. Therefore, approximation (3.67) includes the higher-order multiple scattering effects.

4 Oblique Incidence of Scalar Waves

In this chapter as well as in all our following considerations concerning the transmissivity we will not need any more to assume that the wavefield propagates in a Goupillaud medium. Now, we will derive a set of generalized O'Doherty-Anstey formulas (in the following we call them ODA formulas) for the case of oblique incidence of scalar waves in 1-D randomly inhomogeneous media. These formulas give the attenuation coefficient and the real part of the vertical component of the wave vector of a plane wave transmitted through randomly layered stacks. In fact, in the small-fluctuation approximation, ODA formulas provide a quite comprehensive description of the stratigraphic-filtering effect. They can be directly used in seismic imaging and inversion.

The ODA formulas for scalar waves can be, of course, obtained from the solution for elastic waves given later in the following chapters. However, scalar waves are considered independently in this chapter for the following reasons. First, some of the derivations performed for scalar waves are also needed later for vector elastic waves. Secondly, the range of validity of the ODA formulas seems to be larger for scalar wavefields than for the vector wavefields. This can be shown by numerical experiments, at least.

4.1 Model and Dynamic Equations

We now consider pressure and shear SH-waves in a 1-D medium with density ϱ_{true} and sonic (shear) velocity c_{true} dependent on the vertical coordinate z only. For example, such a medium could represent a stack of many homogeneous layers (see Figure 1) often used to describe sediments. It can be, however, also a continuously heterogeneous medium. Further, for simplicity it is assumed that this vertically heterogeneous medium is embedded between two homogeneous halfspaces with equal constant densities and velocities. These are in turn equal to the spatially averaged density and velocities of the inhomogeneous part of the complete medium. Later we will see that, in spite of this simplification, our solutions for the phase velocity and the attenuation coefficient of the transmissivity are independent of these assumptions.

The uppermost and lowermost boundaries of the inhomogeneous part are $z = 0$ and $z = L$, respectively. The x-axis coincides with the line of intersection of the uppermost interface with the plane of incidence. This plane is normal to the boundaries of the half spaces and to the wavefront of the incident plane wave. The y-axis is then perpendicular to the plane of incidence. A time-harmonic plane wave with circular frequency ω and with the horizontal component k_x of the wave vector impinges from above on the heterogeneous region. The incidence angle is arbitrary but smaller than $\pi/2$. In the heterogeneous part of the medium the depth dependencies of the sonic (shear) velocity c_{true} and density ϱ_{true} are assumed to be typical realizations of random processes. These should be stationary and ergodic. The model is sketched in Figure 1.

In the acoustic case (i.e., for a pressure-wave propagation in a fluid) we need to consider a system of two differential equations for the pressure $p_{true}(\omega, k_x, x, z, t)$ $= \hat{p}(\omega, k_x, z) \exp(ik_x x - i\omega t)$ and the vertical component of the particle velocity $w_{z,true}(\omega, k_x, x, z, t) = w_z(\omega, k_x, z) \exp(ik_x x - i\omega t)$ (Brekhovskikh and Godin, 1990). In the elastic case (shear SH-wave propagation in an isotropic solid) we need to consider a corresponding system of two equations, now however for the displacement velocity u_y and the τ_{yz} stress component (Aki and Richards, 1980). Both systems of equations can be written in the unified form,

$$\frac{\partial}{\partial z} \begin{bmatrix} P_s(z) \\ Q_s(z) \end{bmatrix} = i\omega \begin{bmatrix} 0 & A_s(z) \\ \frac{1}{A_s(z)\omega^2} \left(\frac{\omega^2}{c_{\text{true}}^2(z)} - k_x^2 \right) & 0 \end{bmatrix} \begin{bmatrix} P_s(z) \\ Q_s(z) \end{bmatrix}, \tag{4.1}$$

where we have not indicated the dependence of P_s and Q_s on k_x and ω. To specify this system for sound waves in an inhomogeneous fluid of sonic velocity $c_{\text{true}}(z)$ and density $\varrho_{\text{true}}(z)$, we shall make the following substitutions:

$$P_s(z) = \hat{p}(z), \quad Q_s(z) = w_z(z), \quad A_s(z) = \varrho_{\text{true}}(z). \tag{4.2}$$

For SH-waves, in the corresponding inhomogeneous isotropic elastic medium of shear velocity $c_{\text{true}}(z)$ and density $\varrho_{\text{true}}(z)$, the substitutions have the following form:

$$P_s(z) = u_y(z), \quad Q_s(z) = -\tau_{yz}, \quad A_s(z) = \frac{1}{\varrho_{\text{true}}(z)c_{\text{true}}^2(z)}. \tag{4.3}$$

The variables ω, k_x, x, z and t are independent. They determine the wavefield at any point and time for an incident plane wave defined for $z < 0$ by the horizontal wavenumber k_x and frequency ω. However, to simplify our derivation we define also a homogeneous reference medium with the density ϱ_0 equal to the averaged density of the inhomogeneous medium and the sonic (or shear) velocity c_0 corresponding to the averaged velocity $c_0 = \langle c_{\mathrm{true}} \rangle$ of the inhomogeneous medium. Therefore we introduce the reference wavenumber $k_0 = \omega/c_0$. Consequently, we define the incidence angle ϑ of a plane wave incident in the reference medium by the relation $\sin \vartheta = k_x/k_0$. Thus, in the reference medium the incident plane wave has the following form

$$exp\left[ik_0(z\cos\vartheta + x\sin\vartheta) - i\omega t\right]. \tag{4.4}$$

Furthermore, the density and the sonic (shear) velocity in the inhomogeneous part of the actual medium can be separated in constant parts ϱ_0, c_0 and fluctuating parts ε_ϱ (density fluctuation) and ε_c (velocity fluctuation):

$$c_{\mathrm{true}} = c_0(1 + \varepsilon_c(z)), \qquad \varrho_{\mathrm{true}} = \varrho_0(1 + \varepsilon_\varrho(z)). \tag{4.5}$$

Both fluctuating parts have zero means (due to their definitions). Below they are also referred to simply as 'fluctuations'.

Using the above notations we can rewrite the system (4.1) in the following form:

$$\frac{\partial}{\partial z}\begin{bmatrix} P_1(z) \\ Q_1(z) \end{bmatrix} = \begin{bmatrix} 0 & k_0 M_1(z) \\ -k_0\cos^2\vartheta\, M_2(z) & 0 \end{bmatrix}\begin{bmatrix} P_1(z) \\ Q_1(z) \end{bmatrix} \tag{4.6}$$

with

$$M_1 = 1 + A_1, \quad M_2 = 1 + A_2,$$
$$A_2 = \left[1 + \frac{1 - (1 + \varepsilon_c(z))^2}{\cos^2\vartheta(1 + \varepsilon_c)^2}\right]M_1^{-1} - 1.$$

In system (4.6) we have for acoustic waves

$$P_1 = \frac{\hat{p}(z)}{\varrho_0 c_0^2}, \quad Q_1 = i\frac{w_z(z)}{c_0}, \quad A_1 = \varepsilon_\varrho(z), \tag{4.7}$$

and elastic SH-waves

$$P_1 = \frac{u_y(z)}{c_0}, \quad Q_1 = -i\frac{\tau_{yz}}{\varrho_0 c_0^2}, \quad A_1 = (1 + \varepsilon_c(z))^{-2}(1 + \varepsilon_\varrho(z))^{-1} - 1. \tag{4.8}$$

In the following (after equation (4.12)) we assume that the fluctuations ε_c and ε_ϱ are small. This means that we introduce a small number $\varepsilon \ll 1$ so that $\varepsilon_\varrho(z)$ and $\varepsilon_c(z)$ represent different random stationary processes with the standard deviations being different but as small as $O(\varepsilon)$.

4.2   Strategy of the Solution

Before starting with the derivation of the generalized ODA formulas let us shortly discuss the strategy of the solution. Matrix equation (4.6) is a system of two homogeneous linear first-order differential equations. These must have two linearly independent fundamental solutions (i.e., two unknown column vectors $\mathbf{f_i} = (P_1(z), Q_1(z))_i^T$, with $i = 1, 2$). These form the matrix of fundamental solutions $\mathbf{F_0}(z) = (\mathbf{f_1}(z), \mathbf{f_2}(z))$ which satisfies (Kamke, 1959)

$$\det(\mathbf{F_0}(z)) = \det(\mathbf{F_0}(0)) \exp\left[\int_0^z Tr(\mathbf{Q}(z))dz\right], \qquad (4.9)$$

where $\mathbf{Q}(z)$ denotes the 2×2 matrix on the right-hand side of equation (4.6). However, $Tr(\mathbf{Q}(z))$, i.e., the sum of the diagonal elements of the matrix $\mathbf{Q}$, is equal to zero. Therefore, for all z

$$\det(\mathbf{F_0}(z)) = \det(\mathbf{F_0}(0)) = const. \neq 0, \qquad (4.10)$$

where the last inequality expresses the linear independence of the fundamental solutions.

It is well known that the energy of a scalar wavefield is localized in a 1-D random medium (see Chapter 2). This means that the amplitude of the complex time-harmonic transmissivity T decreases due to multiple scattering like $\exp(-\gamma L)$ for sufficiently large values of L. The attenuation coefficient γ is a positive quantity. This means that matrix equation (4.6) should have one solution decreasing asymptotically (i.e. for L large enough) as $\exp(-\gamma L)$. However, in order to keep the determinant $\det(\mathbf{F_0}(z))$ equal to a non-zero constant, the matrix of the fundamental solutions has to contain another solution that is asymptotically increasing as $\exp(\gamma L)$.

We could repeat all above arguments using the asymptotic properties of products of random matrices (Section 2.6) instead of equations (4.9) and (4.10). For instance, the Virtser's theorem shows that also in the case of the *multimode wave propagation* for any decreasing mode there exists an increasing mode which is characterized by exactly the same attenuation/amplification coefficient.

The same can be said about the phase behavior of the fundamental solutions. There should be one solution with the phase factors $\exp(i\psi L)$ describing the asymptotic behavior of the transmissivity. However, equation (4.10) implies that there should be likewise another solution with the asymptotic behavior of $\exp(-i\psi L)$ describing the up-going wavefield.

The values γ and ψ yield a dynamic equivalent-medium description of the time-harmonic transmissivity when considering its asymptotical behavior for large L. These values are independent of the boundary conditions at $z = 0$ because they characterize the asymptotical behavior of the fundamental solutions.

The general solution of equation (4.6) is a linear combination of two fundamental solutions. Therefore, in case of arbitrary boundary conditions at $z = 0$, the increasing solutions will dominate if L increases. Hence the general solution will behave like the increasing fundamental one. In the search for the solution we will start with convenient boundary conditions at $z = 0$ for the incident pressure- or shear SH-wave. Thereafter, we analyze the behavior of the solutions for large L assuming that the medium fluctuations are small. We will obtain small-perturbation approximations of the quantity γ. This describes the growing exponential factor of the increasing fundamental solution, which is also equal to the attenuation coefficients of the decreasing fundamental solution. The approximation of the quantity ψ will be directly derived.

In order to be convinced that the simple analytic expressions arrived at in the indicated way really describe the transmissivity sufficiently well, we investigate their high- and low- frequency asymptotic behavior and compare the analytic results with those of numerical modeling (see the following two chapters).

4.3   Attenuation and Phase of the Transmissivity

We would like to express the kinematic (ψ) and dynamic (γ) quantities of the transmissivity in terms of a very restricted number of characteristic quantities describing the random medium (e.g., the correlation length l and the variance σ^2 of the medium fluctuations). Therefore, we have to apply a kind of averaging. Different approaches to averaging have been discussed by Stanke and Burridge (1993). These are applied to the transmissivity directly. In contrast to this we apply here averaging to the logarithm of the transmissivity divided by L, i.e., to the attenuation coefficient and to the vertical-phase increment. We know that these are self-averaged quantities (see Chapter 2), which assume their ensemble-averaged values in a single typical realization of the random medium for a large L. Thus, in contrast to the approach of Burridge et al. (1988) we use as the spatial coordinate the depth z rather than the traveltime.

In order to work with the attenuation coefficient and vertical-phase increment of the transmissivity, we have to separate the wavefield into amplitude-like (r_s) and phase-like (ϕ_s) terms:

$$P_1 = r_s \sin \phi_s, \qquad Q_1 = r_s \cos \vartheta \cos \phi_s. \tag{4.11}$$

Using definitions (4.11) we obtain the following differential equations:

$$
\begin{aligned}
\frac{\partial \phi_s}{\partial z} &= \kappa (M_1 \cos^2 \phi_s + M_2 \sin^2 \phi_s) \\
\frac{\partial r_s}{\partial z} &= r_s \frac{\kappa}{2} (M_1 - M_2) \sin 2\phi_s,
\end{aligned}
\tag{4.12}
$$

where $\kappa = k_0 \cos \vartheta$ and the prime denotes the derivative with respect to z.

We now assume that the fluctuations ε_c and ε_ϱ are small. Furthermore, we choose the boundary conditions $\phi_s(0) = 0$ and $r_s(0) = 1$. Then we can solve the system (4.12) by a small-perturbation approximation. The zero-order perturbation approximation is obtained by neglecting the fluctuations. In this case the parameters M_1 and M_2 are equal to 1 (we always suppose that $1/\cos \vartheta$ is finite, i.e., the theory excludes the case $\vartheta \to 90°$). The first equation of the system (4.12) has the approximate zero-order solution $\phi_s^0(z) = \kappa z$. Substitution of this into the right-hand side of the first of equations (4.12) yields a first-order approximation of the phase-like term ϕ_s. Repeating this substitution procedure provides the following approximate solution ϕ_s^{II} for the phase-like term which now accounts for the second order fluctuations:

$$\phi_s^{II} = \kappa z + \kappa \int_0^z dz' \left[A_1(z') \cos^2 \phi_s^I(z') + A_2(z') \sin^2 \phi_s^I(z') \right], \tag{4.13}$$

where

$$\phi_s^I(z) = \kappa z + \kappa \int_0^z dz' \left[A_1(z') \cos^2 \kappa z' + A_2(z') \sin^2 \kappa z' \right] \tag{4.14}$$

It is easy to see that A_1 and A_2 are combinations of first- and second-order small-fluctuation parameters (recall that we neglect terms higher than the second order). They are, therefore, small too. The result (4.14) can be used to write the amplitude-like term accounting for the second order in the small parameters:

$$\ln r_s^{II}(z) = \frac{\kappa}{2} \int_0^z dz' \, [A_1(z') - A_2(z')] \sin(2\phi_s^I).$$ (4.15)

We are now interested in the solution of the posed problem in the limit of thick media, $L \gg \max(l, \lambda)$, where l is the correlation length of the medium and λ is the wavelength in the reference medium. This means that the subject of our study is now related to the limits

$$\psi = \lim_{L \to \infty} \frac{\phi_s(L)}{L}, \quad \gamma = \lim_{L \to \infty} \frac{\ln r_s(L)}{L}.$$ (4.16)

Here we need to evaluate both values at the bottom of the inhomogeneous medium ($z = L$) because at the top ($z = 0$) they have already been defined by the choice of the boundary conditions.

Because of the self–averaging property the limits (4.16) for each typical realization are equal to their expectation values. This allows us to use ensemble averaging for the simplification of equations (4.16). Thus, the averaging in this approach is not a physically crucial step, but it is rather an element of the derivation technique. We consider only terms up to the second order $O(\varepsilon^2)$ in the small parameters. Therefore, we use the approximate solution ϕ_s^{II} for the calculation of γ_1 in equation (4.16). The approximation of $\ln r_s$ in the same order is given in equation (4.15). The further derivation consists of (i) inserting these approximations into the last equations, (ii) expanding the trigonometric functions in Taylor series, considering again only terms not smaller than $O(\varepsilon^2)$, and (iii) calculating the expected values. After these three steps, which are described in details in the appendix to this chapter, we obtain:

$$\psi = \kappa + \frac{\kappa}{2} \langle A_1 + A_2 \rangle - \frac{\kappa^2}{4} \int_0^\infty d\xi \, B_A(\xi) \sin(2\kappa\xi),$$

$$\gamma = \frac{\kappa^2}{4} \int_0^\infty d\xi \, B_A(\xi) \cos(2\kappa\xi).$$ (4.17)

The physical meaning of the coefficients ψ and γ is the following: γ is the *dynamic-equivalent-medium attenuation coefficient* of the transmissivity and ψ is the *dynamic-equivalent medium vertical-phase increment* (i.e., the real part of the vertical component of the wave vector). In the next section these formulas are explicitly written in terms of the correlation functions and the variances of the medium fluctuations. In fact these will then be the final results of the derivation.

In the derivation above we have not used any relationship between L and the small parameter ε. However, we have not studied the convergence of the perturbation series arising from equations similar to (4.13) and (4.15), which describe arbitrary high-order approximations of the vertical phase increment and attenuation coefficient for arbitrary L. If, following Burridge et al. (1988), the restriction $\varepsilon = O(1/\sqrt{L})$ is applied, then the uniform convergence could be proven (see also Section 5.7). However, numerical simulations show that this restriction seems to be too strong for the case of scalar waves.

4.4   Results for Scalar Waves

Substituting the expressions of the quantities A_1 and A_2, given in equations (4.6), (4.7) and (4.8), into equations (4.17) yields the vertical-phase increment ψ_{ac} and the attenuation coefficient γ_{ac} of the pressure-wave transmissivity (i.e., of the scalar acoustic time-harmonic transmissivity)

$$\psi_{ac} = \lambda_{ac} + \omega A_{ac} - \omega^2 \int_0^\infty d\xi\, B_{ac}(\xi)\sin(2\lambda_{ac}\xi),$$
$$\gamma_{ac} = \omega^2 \int_0^\infty d\xi\, B_{ac}(\xi)\cos(2\lambda_{ac}\xi). \tag{4.18}$$

For the case of the scalar elastic SH-wave transmissivity this substitution gives the following formulas for the vertical-phase increment ψ_{SH} and the attenuation coefficient γ_{SH}

$$\psi_{SH} = \lambda_{SH} + \omega A_{SH} - \omega^2 \int_0^\infty d\xi\, B_{SH}(\xi)\sin(2\lambda_{SH}\xi),$$
$$\gamma_{SH} = \omega^2 \int_0^\infty d\xi\, B_{SH}(\xi)\cos(2\lambda_{SH}\xi). \tag{4.19}$$

In equations (4.18) and (4.19) the following new notations are used

$$\lambda_{ac} = \omega\sqrt{a^{-2} - p^2}, \quad \lambda_{SH} = \omega\sqrt{b^{-2} - p^2},$$

$$A_{ac} = \frac{1}{2Xa}\left(X^2\sigma_{\varrho\varrho}^2 + 2\sigma_{a\varrho}^2 + 3\sigma_{aa}^2\right),$$

$$A_{SH} = \frac{1}{2Yb}\left(Y^2\sigma_{\varrho\varrho}^2 + 2\sigma_{b\varrho}^2(2Y^2 - 1) + 4\sigma_{bb}^2(4Y^2 - 1)\right),$$

$$B_{ac} = \frac{1}{X^2a^2}\left(X^4 B_{\varrho\varrho} + 2X^2 B_{a\varrho} + B_{aa}\right),$$

$$B_{SH} = \frac{1}{Y^2b^2}\left(Y^4 B_{\varrho\varrho} + 2B_{b\varrho}Y^2(2Y^2 - 1) + B_{bb}(2Y^2 - 1)^2\right),$$

$$X = \sqrt{1 - a^2p^2}, \quad Y = \sqrt{1 - b^2p^2},$$

(4.20)

where a and b are the averaged velocity c_0 of the pressure wave and the SH-wave, respectively; $p = k_x/\omega$ denotes the horizontal slowness, which is a free parameter of the problem; $B_{\varrho\varrho}(\xi) = \langle\varepsilon_\varrho(z)\varepsilon_\varrho(z+\xi)\rangle$ is the autocorrelation function of the density fluctuations; $B_{aa}(\xi) = \langle\varepsilon_a(z)\varepsilon_a(z+\xi)\rangle$ and $B_{bb}(\xi) = \langle\varepsilon_b(z)\varepsilon_b(z+\xi)\rangle$ are the autocorrelation functions of the sonic-velocity and the shear-velocity fluctuations for the cases of the pressure- and SH-wave, respectively. Here, in the case of the pressure wave the notation ε_a is used for the velocity fluctuation ε_c. In the case of the SH-wave the velocity fluctuation is denoted as ε_b. Using this abbreviations the crosscorrelation functions of the density- and velocity fluctuations are denoted by: $B_{a\varrho}(\xi) = \langle\varepsilon_a(z)\varepsilon_\varrho(z+\xi) + \varepsilon_a(z+\xi)\varepsilon_\varrho(z)\rangle/2$ and $B_{b\varrho}(\xi) = \langle\varepsilon_b(z)\varepsilon_\varrho(z+\xi) + \varepsilon_b(z+\xi)\varepsilon_\varrho(z)\rangle/2$. Finally, the quantities $\sigma_{\varrho\varrho}^2, \sigma_{aa}^2, \sigma_{bb}^2, \sigma_{a\varrho}^2$ and $\sigma_{b\varrho}^2$ denote the corresponding variances, which are equal to the values of the auto- or crosscorrelation functions at the correlation lag $\xi = 0$.

Therefore, the above analysis shows that in the case of $L \gg \max\{l, \lambda\}$ the time-harmonic transmissivity of the pressure wave $T_{ac}(\omega, p, x, z = L, t)$ caused by the incident plane wavefield $T_{ac}(\omega, p, x, z = 0, t)$ is

$$T_{ac}(\omega, p, x, z, t)|_{z=L} = T_{ac}(\omega, p, x, z, t)|_{x,z,t=0} \times$$
$$\times \exp\{i(\psi_{ac}L + px - \omega t) - \gamma_{ac}L\},$$

(4.21)

where the quantity T_{ac} may denote, e.g., the pressure p_{true} or the vertical component of the particle velocity $w_{z,true}$. In the case of the time-harmonic transmissivity of the SH-wave a similar expression is valid:

$$T_{SH}(\omega, p, x, z, t)|_{z=L} = T_{SH}(\omega, p, x, z, t)|_{x,z,t=0} \times$$
$$\times \exp\{i(\psi_{SH}L + px - \omega t) - \gamma_{SH}L\}. \qquad (4.22)$$

Here the quantity T_{SH} may denote e.g., the y-component of the displacement vector.

The exponential terms in equations (4.21) and (4.22) are due to the effect of the stratigraphic filtering. Both equations, together with equations (4.18) and (4.19), constitute the *generalized O'Doherty-Anstey formulas for scalar waves*. They describe the stratigraphic filter. These results show that in order to predict (or correct) the influence of a random stack of layers on an arbitrary planar transmitted wavefield three different correlation functions of the fluctuations of the velocity- and density logs are needed in a general case. The equations (4.18) – (4.22) as shown below reveal a significantly different behavior for both considered cases. Physically this difference can be explained by the fact that a fluid can move independently parallel to an interface, which is impossible in an elastic medium.

4.5   Results in Terms of Reflection-Coefficient-Series Spectra

In the previous chapter we have shown that the original O'Doherty and Anstey (1971) formula was given in terms of the power spectrum of the reflection-coefficient series (see equation 3.32). Sometimes such a representation can also be convenient for the generalized ODA formula.

As actual practical results of log-measurements the discrete series of values $\varrho_{\text{true}}(z_j)$, $a_{\text{true}}(z_j)$ and $b_{\text{true}}(z_j)$ are obtained for $j = 0, \pm 1, \pm 2, \pm 3, \dots$ instead of the continuous functions $\varrho_{\text{true}}(z)$, $a_{\text{true}}(z)$ and $b_{\text{true}}(z)$. The reflection coefficient series is defined then in the following way (we use the reflection coefficients for the displacement):

$$r_j = \frac{\Gamma(z_j) - \Gamma(z_{j+1})}{\Gamma(z_{j+1}) + \Gamma(z_j)}, \qquad (4.23)$$

where the quantity $\Gamma(z)$ is the impedance defined as $\Gamma(z) = \varrho_{\text{true}}(z) a_{\text{true}}(z)$ and $\Gamma(z) = \varrho_{\text{true}}(z) b_{\text{true}}(z)$ for the normally incident P- and S-wave, respectively. Therefore, the quantity r_j is equal to the reflection coefficient of the P- (S-) wave normally incident from the half-space with the parameters $\varrho_{\text{true}}(z_j)$, $a_{\text{true}}(z_j)$ and $b_{\text{true}}(z_j)$ onto the half space with the parameters $\varrho_{\text{true}}(z_{j+1})$, $a_{\text{true}}(z_{j+1})$ and $b_{\text{true}}(z_{j+1})$. The power spectrum of the reflection coefficient series is defined as

$$R_s(\kappa) = \frac{1}{h} \sum_{j=-\infty}^{\infty} e^{-i\kappa j h} \langle r_0 r_j \rangle, \tag{4.24}$$

where $\kappa = \omega/a$ or ω/b, $h = 2\Delta z$ and Δz is the sample interval.

By substituting relations (4.5) into definition (4.23) and neglecting terms of the second and higher order in the medium fluctuations we obtain:

$$r_j \approx \frac{\varepsilon_\Gamma(z_j) - \varepsilon_\Gamma(z_{j+1})}{2} \approx -\frac{h}{4}\frac{\partial \varepsilon_\Gamma(z)}{\partial z}\Big|_{z=z_j}, \tag{4.25}$$

where $\varepsilon_\Gamma(z) = \varepsilon_\varrho(z) + \varepsilon_a(z)$ for the P-wave and $\varepsilon_\Gamma(z) \approx \varepsilon_\varrho(z) + \varepsilon_b(z)$ for the S-wave. Now, the substitution of relation (4.25) into expression (4.24) yields

$$
\begin{aligned}
R_s(\kappa) &= \frac{1}{8} \sum_{j=-\infty}^{j=\infty} \Delta z e^{-i2\kappa j \Delta z} \left\langle \frac{\partial}{\partial z}\varepsilon_\Gamma(z)|_{z=z_0} \frac{\partial}{\partial z}\varepsilon_\Gamma(z)|_{z=z_j} \right\rangle \\
&\approx \frac{1}{8} \int_{-\infty}^{\infty} d\xi e^{-i2\kappa\xi} \left\langle \frac{\partial}{\partial z}\varepsilon_\Gamma(z)|_{z=z_0} \frac{\partial}{\partial(z)}\varepsilon_\Gamma(z)|_{z=z_0+\xi} \right\rangle \\
&= \kappa^2 \int_0^{\infty} d\xi \cos(2\kappa\xi) \langle \varepsilon_\Gamma(0)\varepsilon_\Gamma(\xi) \rangle,
\end{aligned} \tag{4.26}
$$

where the known relationship $\Phi'(\omega) = \omega^2 \Phi(\omega)$ between the power spectrum $\Phi(\omega)$ of a random process and the power spectrum $\Phi'(\omega)$ of the derivative of this process has been applied.

A similar analysis of the relationship between the power spectrum of the reflection coefficient series and the power spectrum of the impedance fluctuations was first given by Banik et al., (1985). This derivation requires the function $\varepsilon_\Gamma(z)$ to be differentiable. A similar derivation, where instead of differentiation finite differences are applied, does not need this restriction. Also an alternative way of the derivation, where the power spectrum of the impedance fluctuations is discretized and reduced to the power spectrum of the reflection coefficient series (Zien, 1993), yields the same result without any assumptions about the smoothness of the function $\varepsilon_\Gamma(z)$.

Using the definition of the function $\varepsilon_\Gamma(z)$ (see the line below equation 4.25) we obtain from equation (4.26) the following relation

$$R_s(\kappa) = \kappa^2 \int_0^\infty d\xi cos(2\kappa\xi) \left[B_{\varrho\varrho}(\xi) + 2B_{a\varrho,b\varrho}(\xi) + B_{aa,bb}(\xi)\right]. \tag{4.27}$$

Comparing result (4.27) with the formulas for the attenuation coefficients of the P-, SV- and SH-wave (see equations (4.18), (4.19), (5.36) and (5.37)) and assuming that the absolute value of the incident wavefield is equal to one, a series of formulas can be found for the absolute value of the transmissivity:

- In the case of vertical incidence these formulas have the form of the classical ODA formula, which is valid for the pressure wave in an acoustic medium as well as for the SV-, SH- and P-waves in an elastic medium:

$$|T| = \exp\left(-LR_s(\kappa)\right). \tag{4.28}$$

- For the pressure wave, in an acoustic medium with constant density, and for the P-wave in an elastic medium with constant density and constant shear modulus we obtain

$$|T| = \exp\left(-LR_s(\lambda_{ac})\frac{1}{X^4}\right) = \exp\left(-LR_s(\kappa\cos\vartheta)\frac{1}{\cos^4\vartheta}\right), \tag{4.29}$$

where the angle of incidence ϑ is defined for the homogeneous reference medium as $\sin\vartheta = pa$, and $\kappa = \omega/a$.
- For the SH-wave in an elastic medium with constant density the absolute value of the transmissivity is

$$|T| = \exp[-LR_s(\lambda_{SH})(2 - \frac{1}{Y^2})^2] = \exp[-LR_s(\kappa\cos\vartheta)(2 - \frac{1}{\cos^2\vartheta})^2] \tag{4.30}$$

where $\sin\vartheta = pb$ and $\kappa = \omega/b$.

Both formulas (4.29) and (4.30) show that usually (e.g., in the random medium with exponential correlation functions) the attenuation increases with the angle of incidence for the compressional wave, whereas for the SH-wave it decreases to zero as the angle approaches 45°. Thereafter it increases again with the angle. This behavior of the attenuation in the neighborhood of 45° leads to an almost total absence of dispersion of SH-waves for such angles. This is the *Brewster anomaly effect* for SH-wave also known for the *p*-polarized light (Sipe et al., 1988).

In the general case of oblique incidence of a plane wave onto a layered stack with fluctuating densities and velocities it is not convenient to express the generalized O'Doherty-Anstey formulas in terms of the spectrum $R(\kappa)$. The reason is that one runs into some problems when using this particular parameterization of the medium to express the above results for the vertical-phase increments.

4.6   Appendix: Details of the Derivation for ψ and γ

In this appendix we show details of the derivation of equations (4.17). Inserting approximations (4.13) and (4.15) into limits (4.16) gives

$$
\psi = \lim_{L \to \infty} \frac{1}{L} \{ \kappa L + \frac{\kappa}{2} \langle A_1 + A_2 \rangle L - \frac{\kappa}{2} \langle A_2 - A_1 \rangle \int_0^L dz \cos(2\kappa z)
$$

$$
+ \frac{\kappa^2}{2} \int_0^L dz \int_0^z dz' C_A(z', z) \sin(2\kappa z)
$$

$$
- \frac{\kappa^2}{2} \int_0^L dz \int_0^z dz' B_A(z', z) \sin(2\kappa z) \cos(2\kappa z') \}
$$

$$
\gamma = \lim_{L \to \infty} \frac{1}{L} \{ -\frac{\kappa}{2} \langle A_2 - A_1 \rangle \int_0^L dz \sin(2\kappa z)
$$

$$
- \frac{\kappa^2}{2} \int_0^L dz \int_0^z dz' C_A(z', z) \cos(2\kappa z)
$$

$$
+ \frac{\kappa^2}{2} \int_0^L dz \int_0^z dz' B_A(z', z) \cos(2\kappa z) \cos(2\kappa z') \},
$$

(4.31)

where the quantities

$$
B_A(z', z) = \langle (A_2(z') - A_1(z'))(A_2(z) - A_1(z)) \rangle ,
$$
$$
C_A(z', z) = \langle (A_2(z') + A_1(z'))(A_2(z) - A_1(z)) \rangle
$$

are combinations of the crosscorrelation and autocorrelation functions of ε_c and ε_ϱ (in the expressions for B_A and C_A, besides the averaging, we also must neglect terms which are smaller than $O(\varepsilon^2)$). In the case of a stationary randomly inhomogeneous medium the quantities B_A and C_A are functions of the correlation lag, $\xi = z' - z$, only. Furthermore, the quantities $\langle A_2(z) - A_1(z) \rangle$ and $\langle A_2(z) + A_1(z) \rangle$ are constants. Taking this into account we obtain:

$$\psi = \lim_{L \to \infty} \frac{1}{L} \{ \kappa L + \frac{\kappa}{2} \langle A_1 + A_2 \rangle L - \frac{1}{4} \langle A_2 - A_1 \rangle \sin(2\kappa L)$$

$$+ \frac{\kappa}{2} \int_0^L \sin(2\kappa z) C_{int}(z) dz$$

$$- \frac{\kappa^2}{4} \int_0^L dz \int_0^z dz' B_A(\xi) \left[\sin(2\kappa(z' + z)) - \sin(2\kappa(\xi)) \right] \}$$

$$\gamma = \lim_{L \to \infty} \frac{1}{L} \{ \frac{1}{4} \langle A_2 - A_1 \rangle \left(\cos(2\kappa L) - 1 \right)$$

$$- \frac{\kappa^2}{2} \int_0^L \cos(2\kappa z) C_{int}(z) dz$$

$$+ \frac{\kappa^2}{4} \int_0^L dz \int_0^z dz' B_A(\xi) \left[\cos(2\kappa(z' + z)) + \cos(2\kappa(\xi)) \right] \},$$

(4.32)

where $C_{int} = \int_{-z}^0 C_A(\xi) d\xi$. It is not very difficult to see that the third term in the expression for ψ can be neglected in comparison with the first two terms increasing with L as $O(L)$. The same can be said about the first term in γ (it is also negligible in comparison with terms of the order $O(L)$). In order to consider the other terms let us introduce the new integration variables:

$$\xi = z' - z, \quad \eta = (z' + z)/2. \tag{4.33}$$

Because of this the range of integration has been changed from the old one

$$0 \leq z \leq L, \quad 0 \leq z' \leq z \tag{4.34}$$

to the new one (see also Figure 8)

$$-L \leq \xi \leq 0, \quad -\frac{\xi}{2} \leq \eta \leq L + \frac{\xi}{2}. \tag{4.35}$$

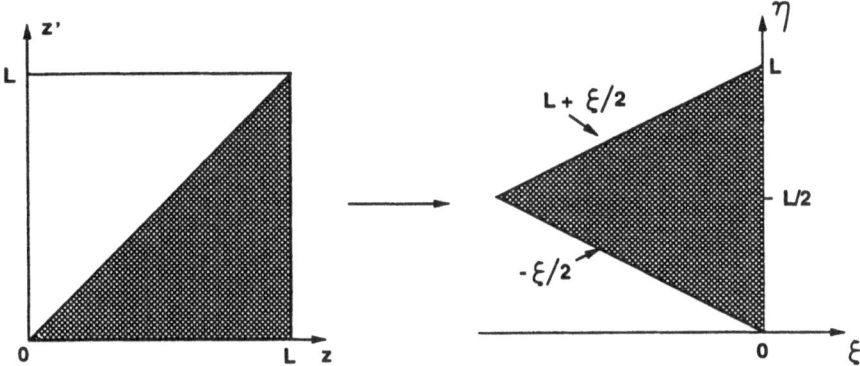

Fig. 8. Range of integration for z and z' and the corresponding range for ξ and η.

Thus, the transformation of coordinates yields

$$\psi = \lim_{L\to\infty}\frac{1}{L}\{\kappa L + \frac{\kappa}{2}\langle A_1 + A_2\rangle L + L\frac{\kappa^2}{4}\int_{-L}^{0} d\xi\, B_A(\xi)\sin(2\kappa(\xi))$$

$$+ \frac{\kappa}{4}\int_{-L}^{0} d\xi[(\cos(2\kappa\xi) - \cos(2\kappa L))\, C_A(\xi)$$

$$+ \kappa B_A(\xi)\,(\cos(4\kappa L + 2\kappa\xi)/4 - \cos(2\kappa\xi)/4 + \xi\sin(2\kappa(\xi)))]\}$$

$$(4.36)$$

$$\gamma = \lim_{L\to\infty}\frac{1}{L}\{L\frac{\kappa^2}{4}\int_{-L}^{0} d\xi\, B_A(\xi)\cos(2\kappa(\xi))$$

$$- \frac{\kappa}{4}\int_{-L}^{0} d\xi[(\sin(2\kappa\xi) - \sin(2\kappa L))\, C_A(\xi)$$

$$+ \kappa B_A(\xi)\,(\sin(4\kappa L + 2\kappa\xi)/4 + \sin(2\kappa\xi)/4 + \xi\cos(2\kappa(\xi)))]\}.$$

In most realistic media the absolute values of the functions $B_A(\xi)$ and $C_A(\xi)$ quickly vanish outside a finite domain $|\xi| \leq l$ of the correlation lag, ξ. This is due to the loss of correlation with increasing ξ. The integrals converge quickly. Therefore, the last integral terms in the expressions for ψ and γ can be neglected in comparison with the terms of the order of $O(L)$. Another property of the functions $B_A(\xi)$ and $C_A(\xi)$ is also important for further simplifications: they are even functions of their argument. Now, taking the limits we finally obtain equations (4.17).

5 Elastic P-SV Waves

In the two previous chapters the propagation of scalar waves has been studied. Only a single-mode wave propagation was considered. Scattering did not change the mode of the wave, i.e., pressure waves in a fluid are scattered into pressure waves again. The same can also be said about SH-waves. However, mode conversion takes place for elastic P- and SV-wavefields. In the process of their propagation in a multilayered medium, mode conversion takes place at any interface. Therefore, physically, the situation in elastic media is qualitatively different: it is a *multimode wave propagation*.

Quantitative consequences of the difference between single- and multimode wave propagation will be shown in this chapter, where we study the influence of 1-D randomly stratified media on the *transmissivity* of obliquely incident elastic plane waves. For this a combination of the *invariant embedding method* with the *small-perturbation expansion* is used. The resulting analytic expressions for the plane time-harmonic and transient P- and SV-waves transmissivities and reflectivities can be used for both deterministically and statistically specified stratifications. They do not have any essential restrictions concerning the frequency range of their validity. However, it is assumed that the fluctuations of the density and compressional and shear velocities are small in comparison to the average values of these parameters. The analytic expressions for the attenuation coefficients and phase velocities of the P- and SV transmissivities predict well their smoothed behavior in one single realization of a 1-D random medium. Both expressions can be considered as fundamental wave attributes for a *dynamic-equivalent medium*. They can also be looked upon as a generalization of the ODA formula, which can account now for oblique incidence and mode-conversion effects in an inhomogeneous 1-D medium with depth-dependent (i.e., non-stationary) statistics of fluctuations. Likewise the resulting frequency-dependent phase velocities can be considered as a broad-frequency-range generalization of their static (i.e., low-frequency) approximations, obtained by what is known as *Backus averaging* (see Backus, 1962). They postulate a *frequency-dependent elastic transverse isotropy* of the dynamic-equivalent medium, which replaces the 1-D random medium.

As mentioned above the derivation in this chapter is based on the invariant-

embedding method, which reduces a boundary-value problem to a problem with
initial conditions. The invariant-embedding method will be combined with the
second-order Rytov small-perturbation approximation for elastic waves. In order
to study the transmissivity we apply the perturbation technique directly to its
logarithm rather than to the transmissivity itself. The statistical averaging of
such logarithms in random stationary media allows us to obtain fairly simple
analytical expressions for both the attenuation coefficient and vertical slowness
of the transmissivity. Such an averaging is justified for single realizations of
a random medium due to the self-averaging property of both quantities. Self-
averaging as well as wavefield localization follows from the asymptotic properties
of products of a large number of random matrices (i.e., the propagator matrices;
see Chapter 2).

In the following sections we provide a description of the earth model and the
dynamic equations considered in this chapter. We parameterize the inhomogeneous 1-D medium by introducing small fluctuations to the homogeneous reference medium. These fluctuations can be either deterministic or random with no
constraints on their statistics. All derivations until equations (5.35) are valid for
both cases.

After introducing the general dynamic equations we perform the eigenvalue decomposition of their coefficient matrix for the case of vanishing fluctuations.
Then we transform the general dynamic equations in such a way that this matrix
becomes diagonal. After this we derive the system of the invariant-embedding
equations for the reflectivity and transmissivity. By solving this system with the
small-perturbation approximation we arrive at analytical approximations for the
transmissivity.

5.1   Model and Dynamic Equations

Let us consider the transmissivity and reflectivity caused by either a plane P- and
SV-wave obliquely incident from an upper homogeneous half-space on a thick
1-D inhomogeneous elastic medium embedded between two equal homogeneous
half spaces (see Figure 1). The density ϱ_{true} and the compressional- and shear-
wave velocities a_{true} and b_{true} depend on the depth coordinate z only. Both
homogeneous half-spaces have constant density and velocities, which are equal
to the spatially averaged density and velocities of the inhomogeneous medium.

We assume that in the inhomogeneous medium, within the uppermost and lowermost boundaries $z = 0$ and $z = L$, the velocity and density distributions can be
looked upon as deterministic functions of z or as realizations of random processes.
It is expected that the fluctuations of all three physical medium parameters

are relatively small if compared to their spatially averaged values ($\approx$ 20% or smaller). We do not suppose that the functions $\varrho_{\text{true}}(z)$, $a_{\text{true}}(z)$ and $b_{\text{true}}(z)$ are differentiable nor piecewise differentiable. In fact, we not even require that they are stationary.

The x-axis coincides with the line of intersection of the uppermost interface with the plane of incidence (such a plane is normal to the boundaries of both half spaces and to the wavefront of the incident plane wave). The y-axis is then perpendicular to the plane of incidence. From the uppermost homogeneous half-space ($z < 0$) either a plane P- or SV-wave with circular frequency ω and with the horizontal component $k_x = \omega p$ (p is the horizontal slowness) of the wave vector impinges on the plane $z = 0$. As we consider the S-wave to be polarized in the vertical incidence plane we refer to it as the SV-wave. With this choice of the coordinate system the SH-wave is polarized along the y-axis, i.e. perpendicular to the plane of incidence. It can thus be considered as a scalar wave and as such it was already treated in the previous chapter.

It is now our aim to analytically approximate in a simple way

(a) the time-harmonic P and SV reflectivity (i.e., the reflected response in the uppermost half space to either an obliquely downgoing time-harmonic P- or SV plane wave, incident in the same half space) and

(b) the vertical-phase increment and attenuation coefficient, which both define the time-harmonic P and SV transmissivity (i.e., the response in the lowermost half space to either an oblique time-harmonic P or SV plane wave, incident in the uppermost half space).

Let us write the P-SV wavefield in the inhomogeneous medium as the vector $\mathbf{f}(z)\exp[i\omega(px - t)]$, where t is the time and $\mathbf{f}(z) \equiv (u_x(z), u_z(z), \tau_{zx}(z), \tau_{zz}(z))^T$ is a column vector consisting of the indicated components of the displacement vector $\mathbf{u}$ and stress tensor τ at any depth for which $x = 0$ and at $t = 0$. The column vector $\mathbf{f}(z)$ satisfies the following first-order differential matrix equation (Aki, Richards, 1980, pp.163-167, 267-270):

$$\frac{\partial}{\partial z}\mathbf{f}(z) = \mathbf{Q}(z)\mathbf{f}(z), \tag{5.1}$$

where

$$\mathbf{Q}(z) = \begin{pmatrix} 0 & -i\omega p & \varrho_{\text{true}}^{-1}(z)b_{\text{true}}^{-2}(z) & 0 \\ -i\omega p(1 - 2\nu(z)) & 0 & 0 & \varrho_{\text{true}}^{-1}(z)a_{\text{true}}^{-2}(z) \\ \omega^2\varrho_{\text{true}}(z)q(z) & 0 & 0 & -i\omega p(1 - 2\nu(z)) \\ 0 & -\varrho_{\text{true}}(z)\omega^2 & -i\omega p & 0 \end{pmatrix} \tag{5.2}$$

with $q(z) = 4p^2 b_{\text{true}}^2(z)(1 - \nu(z)) - 1$ and $\nu(z) = b_{\text{true}}^2(z)/a_{\text{true}}^2(z)$. The elements of matrix $\mathbf{Q}(z)$ are combinations of the density, the velocities of compressional and shear waves, the horizontal slowness p and frequency ω. The density and the velocities of the compressional and shear waves in the inhomogeneous region $0 \leq z \leq L$ of the medium can be separated into the constant parts ϱ, a and b and the corresponding depth-dependent parts $\varrho \varepsilon_\varrho(z)$, $a\varepsilon_a(z)$ and $b\varepsilon_b(z)$:

$$
\begin{aligned}
\varrho_{\text{true}}(z) &= \varrho(1 + \varepsilon_\varrho(z)), \\
a_{\text{true}}(z) &= a(1 + \varepsilon_a(z)), \\
b_{\text{true}}(z) &= b(1 + \varepsilon_b(z)).
\end{aligned}
\tag{5.3}
$$

The quantities ϱ, a and b, therefore, characterize the *homogeneous reference medium*, which can be defined in different ways. We define the quantities ϱ, a and b as the following spatial averages

$$
\begin{aligned}
\varrho &= \frac{1}{L} \int_0^L \varrho_{\text{true}}(z)dz, \\
a &= \frac{1}{L} \int_0^L a_{\text{true}}(z)dz, \\
b &= \frac{1}{L} \int_0^L b_{\text{true}}(z)dz.
\end{aligned}
\tag{5.4}
$$

All three depth-dependent quantities $\varepsilon_\varrho(z)$, $\varepsilon_a(z)$ and $\varepsilon_b(z)$ have zero means (due to their definitions). In spite of their arbitrary (i.e., deterministic or random) character, they are referred to below as the 'medium fluctuations' or simply as the 'fluctuations'. Later we suppose that the fluctuations are small. This means that we introduce a small number $\varepsilon \ll 1$ so that $\varepsilon_\varrho(z), \varepsilon_a(z)$ and $\varepsilon_b(z)$ are of the order of $O(\varepsilon)$.

5.2   Transformation of the Dynamic Equations

The representation (5.3) of the physical medium parameters leads to a new form of equation (5.1):

$$
\frac{\partial}{\partial z}\mathbf{f}(z) = \mathbf{Q}_0 \mathbf{f}(z) + \mathbf{Q}_\varepsilon(z)\mathbf{f}(z),
\tag{5.5}
$$

where we define $\mathbf{Q_0}$ as the following matrix independent of z

$$
\mathbf{Q_0} \equiv \begin{pmatrix}
0 & -i\omega p\, \varrho^{-1}b^{-2} & & 0 \\
-i\omega p(a^2 - 2b^2)a^{-2} & 0 & 0 & \varrho^{-1}a^{-2} \\
\omega^2\varrho(4p^2b^2(a^2 - b^2)a^{-2} - 1) & 0 & 0 & -i\omega p(a^2 - 2b^2)a^{-2} \\
0 & -\varrho\omega^2 & -i\omega p & 0
\end{pmatrix}
\tag{5.6}
$$

and $\mathbf{Q_\epsilon} \equiv \mathbf{Q} - \mathbf{Q_0}$ is the *fluctuation matrix*

$$
\mathbf{Q_\epsilon}(z) = \begin{pmatrix}
0 & 0 & \varepsilon_{13}(z) & 0 \\
i\varepsilon_{21}(z) & 0 & 0 & \varepsilon_{24}(z) \\
\varepsilon_{31}(z) & 0 & 0 & i\varepsilon_{21}(z) \\
0 & \varepsilon_{42}(z) & 0 & 0
\end{pmatrix} .
\tag{5.7}
$$

The five real quantities $\varepsilon_{ij}(z)$ are combinations of terms with first-order and higher-order powers in the fluctuations (see Appendix 5.8).

If the fluctuations are zero ($\mathbf{Q_\epsilon} = 0$) then equation (5.5) will describe the P-SV wave field in the homogeneous reference medium, which is characterized by the matrix $\mathbf{Q_0}$. In this case equation (5.5) can be linearly transformed to four uncoupled equations describing four independent up- and downgoing plane P- and S-waves. After this transformation the matrix $\mathbf{Q_0}$ becomes diagonal.

In order to diagonalize the matrix $\mathbf{Q_0}$ its *eigenvalue decomposition* must be performed. The eigenvalues of the matrix $\mathbf{Q_0}$ are $i\kappa_a$, $i\kappa_b$, $-i\kappa_a$ and $-i\kappa_b$, where $\kappa_a = \omega\sqrt{(1/a^2 - p^2)}$ and $\kappa_b = \omega\sqrt{(1/b^2 - p^2)}$. The four eigenvectors of the matrix $\mathbf{Q_0}$ serve as columns of the eigenvector matrix $\mathbf{E_0}$. This and its inverse matrix $\mathbf{E_0}^{-1}$ can be found in Aki and Richards (1980, formulas (5.65) and (5.67), pp. 166-167).

We can now transform the unknown column vector $\mathbf{f}$ into another unknown column vector $\mathbf{g}(z) \equiv (d_p(z), d_s(z), u_p(z), u_s(z))^T$:

$$
\mathbf{g}(z) = \mathbf{E_0}^{-1}\mathbf{f}(z).
\tag{5.8}
$$

The elements of $\mathbf{g}(z)$ are linear combinations of the displacement and stress components $u_x(z)$, $u_z(z)$, $\tau_{zx}(z)$, and $\tau_{zz}(z)$. It can be seen that in the absence of fluctuations (i.e., within the homogeneous reference medium) the components $d_p(z), d_s(z), u_p(z), u_s(z)$ are just the displacements in the down- and upgoing P- and SV- waves, respectively (the notations are obvious). Strictly speaking, these are the projections of the displacement vectors onto the polarization directions of the corresponding waves at the point $(t = 0, x = 0, z = 0)$. Obviously we also have

$$\mathbf{f}(z) = \mathbf{E}_0 \mathbf{g}(z). \tag{5.9}$$

Using this latter relation and equation (5.5) we obtain the following form of equation (5.1)

$$\frac{\partial}{\partial z}\mathbf{g}(z) = \Lambda_0 \mathbf{g}(z) + \Lambda_\epsilon(z)\mathbf{g}(z), \tag{5.10}$$

where

$$\Lambda_0 = \mathbf{E}_0^{-1}\mathbf{Q}_0\mathbf{E}_0, \tag{5.11}$$

and

$$\Lambda_\epsilon(z) = \mathbf{E}_0^{-1}\mathbf{Q}_\epsilon(z)\mathbf{E}_0. \tag{5.12}$$

It follows from equation (5.11) that matrix Λ_0 is the diagonal 4×4 matrix

$$\Lambda_0 = diag[i\kappa_a, i\kappa_b, -i\kappa_a, -i\kappa_b]. \tag{5.13}$$

Equation (5.12) explicitly yields the following matrix $\Lambda_\epsilon(z)$:

$$\Lambda_\epsilon(z) = \begin{pmatrix} A(z) & B(z) & C(z) & D(z) \\ E(z) & F(z) & G(z) & H(z) \\ -C(z) & -D(z) & -A(z) & -B(z) \\ -G(z) & -H(z) & -E(z) & -F(z) \end{pmatrix}. \tag{5.14}$$

The quantities A, B, C, D, E, F, G and H are combinations of terms of first-order and higher-order powers in the fluctuations (see Appendix 5.8).

5.3   Invariant Embedding

From equation (5.10) we obtain:

$$\frac{\partial}{\partial z}\mathbf{g}(z) = \mathbf{M}(z)\mathbf{g}(z), \tag{5.15}$$

where the matrix $\mathbf{M} = \Lambda_0 + \Lambda_\varepsilon(z)$:

$$\mathbf{M} = \begin{pmatrix} i\kappa_a + A(z) & B(z) & C(z) & D(z) \\[2mm] E(z) & i\kappa_b + F(z) & G(z) & H(z) \\[2mm] -C(z) & -D(z) & -i\kappa_a - A(z) & -B(z) \\[2mm] -G(z) & -H(z) & -E(z) & -i\kappa_b - F(z) \end{pmatrix}. \tag{5.16}$$

Now, let us introduce the 4×4 propagator matrix $\mathbf{P}(\zeta, 0)$. This links the solutions $\mathbf{g}(z)$ at the two depths $z = 0$ and $z = \zeta$ (for a more thorough description of the concept of a propagator matrix see Kennett, 1983):

$$\mathbf{g}(\zeta) = \mathbf{P}(\zeta, 0)\mathbf{g}(0). \tag{5.17}$$

Gilbert and Backus (1966) showed that $\mathbf{P}(\zeta, 0)$ satisfies the equation

$$\frac{\partial}{\partial \zeta}\mathbf{P}(\zeta, 0) = \mathbf{M}(\zeta)\mathbf{P}(\zeta, 0). \tag{5.18}$$

Let us imagine that the depth of the lower bound of the inhomogeneous medium is the variable ζ and let us look for a solution of the problem for $\zeta = L$. Then the solution $\mathbf{P}(\zeta, 0)$ of equation (5.18) with the initial condition $\mathbf{P}(0, 0) = \mathbf{I}$ ($\mathbf{I}$ is the identity matrix) is according to equation (5.17) a propagator matrix. This links the vectors $\mathbf{g}(0)$ and $\mathbf{g}(\zeta)$ at the top $z = 0$ and at the bottom $z = \zeta$ of the inhomogeneous part of the medium. For $z < 0$ and $z > \zeta$ we have the elastic parameters of the homogeneous reference medium. In the homogeneous reference medium (including the planes $z = 0$ and $z = \zeta$) the vector $\mathbf{g}(z)$ can be rewritten as follows

$$\mathbf{g}(z) = \begin{pmatrix} \mathbf{d}(z) \\ \mathbf{u}(z) \end{pmatrix}, \tag{5.19}$$

with the vectors $\mathbf{d}(z) = (d_p(z), d_s(z))^T$ and $\mathbf{u}(z) = (u_p(z), u_s(z))^T$ representing the downgoing and the upgoing waves. Substituting $\mathbf{g}(0)$ and $\mathbf{g}(\zeta)$ in the form (5.19) into equation (5.17) and taking *the radiation condition* (there are no waves coming from infinity apart from either the incident downgoing P- or SV-wave) into account, it is straightforward to obtain

$$\mathbf{P}(\zeta, 0) = \begin{pmatrix} \mathbf{T_d}(\zeta, 0) - \mathbf{R_u}(\zeta, 0)\mathbf{T_u}^{-1}(\zeta, 0)\mathbf{R_d}(\zeta, 0) & \mathbf{R_u}(\zeta, 0)\mathbf{T_u}^{-1}(\zeta, 0) \\ -\mathbf{T_u}^{-1}(\zeta, 0)\mathbf{R_d}(\zeta, 0) & \mathbf{T_u}^{-1}(\zeta, 0) \end{pmatrix} \tag{5.20}$$

where the 2×2 submatrices $\mathbf{T_d}(\zeta, 0), \mathbf{R_d}(\zeta, 0), \mathbf{R_u}(\zeta, 0)$ and $\mathbf{T_u}(\zeta, 0)$ satisfy the following equations

$$\begin{aligned} \mathbf{d}(\zeta) &= \mathbf{T_d}(\zeta, 0)\mathbf{d}(0), \quad \mathbf{u}(0) = \mathbf{R_d}(\zeta, 0)\mathbf{d}(0), \\ \mathbf{u}(0) &= \mathbf{T_u}(\zeta, 0)\mathbf{u}(\zeta), \quad \mathbf{d}(\zeta) = \mathbf{R_u}(\zeta, 0)\mathbf{u}(\zeta). \end{aligned} \tag{5.21}$$

The physical meaning of the four elements of these submatrices becomes clear from the substitution of the vectors $\mathbf{d}$ and $\mathbf{u}$ in form of incident and transmitted/reflected waves into equations (5.21). For instance, if $\mathbf{d}(0) = (0, 1)^T$ then $\mathbf{d}(\zeta) = (t_{dps}, t_{dss})^T$ and $\mathbf{u}(0) = (r_{dps}, r_{dss})^T$, where the quantities t_{dps} and t_{dss} are the transmissivities describing the transmitted P- and SV-wave responses, respectively, at the depth ζ due to the incident *downgoing* SV-wave at the interface $z = 0$. Correspondingly, the quantities r_{dps} and r_{dss} are the reflectivities describing the corresponding P- and SV-wave responses reflected at the interface $z = 0$. Such substitutions yield:

$$\mathbf{R_d} = \begin{pmatrix} r_{dpp} & r_{dps} \\ r_{dsp} & r_{dss} \end{pmatrix}, \quad \mathbf{T_d} = \begin{pmatrix} t_{dpp} & t_{dps} \\ t_{dsp} & t_{dss} \end{pmatrix},$$

$$\mathbf{R_u} = \begin{pmatrix} r_{upp} & r_{ups} \\ r_{usp} & r_{uss} \end{pmatrix}, \quad \mathbf{T_u} = \begin{pmatrix} t_{upp} & t_{ups} \\ t_{usp} & t_{uss} \end{pmatrix}, \tag{5.22}$$

where the quantities $t_{dpp}, t_{dsp}, r_{dpp}$ and r_{dsp} are the corresponding transmissivities and reflectivities for the incident *downgoing* P-wave (at the interface $z = 0$) assuming an inhomogeneous 1-D medium confined to $0 < z < \zeta$. Similarly the quantities $t_{upp}, t_{usp}, r_{upp}, r_{usp}, t_{ups}, t_{uss}, r_{ups}$ and r_{uss} describe the transmitted and reflected P- and SV-waves for *upgoing* P- and SV-waves incident at the interface $z = \zeta$. Therefore, the propagator matrix $\mathbf{P}(\zeta, 0)$ has the partitioned form

(5.20), where the elements of $\mathbf{T_d}(\zeta,0), \mathbf{R_d}(\zeta,0), \mathbf{R_u}(\zeta,0)$ and $\mathbf{T_u}(\zeta,0)$ encompass the respective transmissivities and reflectivities.

By substituting the partitioned form (5.20) of the matrix $\mathbf{P}(\zeta,0)$ into equation (5.18) we arrive at the system of the invariant-embedding equations:

$$\frac{\partial}{\partial\zeta}\mathbf{R_u} = \mathbf{M_2} + \mathbf{M_1}\mathbf{R_u} + \mathbf{R_u}\mathbf{M_1} + \mathbf{R_u}\mathbf{M_2}\mathbf{R_u},$$

$$\frac{\partial}{\partial\zeta}\mathbf{T_u} = \mathbf{T_u}\mathbf{M_1} + \mathbf{T_u}\mathbf{M_2}\mathbf{R_u},$$

$$\frac{\partial}{\partial\zeta}\mathbf{T_d} = \mathbf{M_1}\mathbf{T_d} + \mathbf{R_u}\mathbf{M_2}\mathbf{T_d}, \tag{5.23}$$

$$\frac{\partial}{\partial\zeta}\mathbf{R_d} = \mathbf{T_u}\mathbf{M_2}\mathbf{T_d},$$

where

$$\mathbf{M_1} = \begin{pmatrix} i\kappa_a + A(\zeta) & B(\zeta), \\ E(\zeta) & i\kappa_b + F(\zeta) \end{pmatrix},$$

$$\mathbf{M_2} = \begin{pmatrix} C(\zeta)\ D(\zeta) \\ G(\zeta)\ H(\zeta) \end{pmatrix}. \tag{5.24}$$

The first equation of system (5.23) is a matrix Riccati equation for the submatrix $\mathbf{R_u}$. It is an independent equation and must be solved first. Thereafter the other three equations can be solved step by step. If the submatrices satisfy the initial conditions

$$\mathbf{T_u}(\zeta = 0,0) = \mathbf{I}, \quad \mathbf{T_d}(\zeta = 0,0) = \mathbf{I},$$
$$\mathbf{R_u}(\zeta = 0,0) = \mathbf{0}, \quad \mathbf{R_d}(\zeta = 0,0) = \mathbf{0}, \tag{5.25}$$

with $\mathbf{I}$ and $\mathbf{0}$ being the 2×2 identity and zero matrices, then the elements of the matrices (5.22), obtained by integrating equations (5.23) from $\zeta = 0$ to $\zeta = L$, will provide the searched-for reflectivities and transmissivities.

System (5.23) is a system of 16 coupled ordinary differential equations for the elements of the submatrices $\mathbf{T_u}, \mathbf{T_d}, \mathbf{R_u}$ and $\mathbf{R_d}$. It can serve as a general basis for an analysis of the transmissivities and reflectivities of elastic waves. For instance, a low-frequency or small-fluctuation approximation can be developed from it in a straightforward way. As indicated above, this system of equations is valid for an arbitrary statistic as well as deterministic parameterization of the 1-D inhomogeneous medium.

5.4   Transmissivity in Case of Small Fluctuations

Below we assume the medium fluctuations to be arbitrary but small. For convenience we write the transmissivities as

$$
\begin{aligned}
t_{upp} &= e^{i\kappa_a \zeta + \Psi_{up}}, \quad t_{uss} = e^{i\kappa_b \zeta + \Psi_{us}}, \\
t_{dpp} &= e^{i\kappa_a \zeta + \Psi_{dp}}, \quad t_{dss} = e^{i\kappa_b \zeta + \Psi_{ds}}.
\end{aligned}
\tag{5.26}
$$

Thus the derivation of the transmissivities is reduced to finding the complex phase Ψ. This leads to their *Rytov approximations*. In the following we derive the second Rytov approximations of the transmissivities, which lead to the generalized O'Doherty-Anstey formulas. In 1-D heterogeneous media Rytov approximations take back-scattering as well as forward-scattering effects into account. This is in contrast to the case of 2-D and 3-D heterogeneous media, where the back scattering is usually neglected (see, e.g., Shapiro and Kneib, 1993 and Shapiro, Schwarz and Golg, 1996). Moreover, for transmitted wavefields the Rytov approximations seem to be more accurate than the corresponding *Born approximations* (see e.g., Ishimaru, 1978, Chapter 17). This is due to the fact that exponential representations, like equations (5.26), implicitly take more scattering effects into account than some first terms of the Born series.

Substituting definitions (5.26) into system (5.23) and taking the initial conditions (5.25) into account, we conclude that all unknown quantities in this system are of the order $O(\varepsilon)$. In the following we expand the exponential terms into a Taylor series and neglect all terms of the third and higher orders ($O(\varepsilon^n)$; $n \geq 3$). This is, however, not always allowed. For instance, in the case of random fluctuations in the limit of a semi-infinite inhomogeneous medium (i.e., $L \longrightarrow \infty$) the absolute values of the P-P and SV-SV reflectivities can tend to limits of the order of 1 due to the localization. Also the real parts of Ψ tend to infinity. The higher-order terms can be neglected if the thickness L of the inhomogeneous medium satisfies the inequality $L < max(\lambda, l)/\varepsilon^2$, where λ is a wavelength of the considered wave and l is a characteristic size of inhomogeneities (e.g., their correlation length). In fact, this situation can be called a *weak localization*.

For the analysis of the transmissivities of the P- and SV-waves caused by an incident plane wave in the uppermost homogeneous half space we need only eight equations. After performing the above mentioned Taylor expansion and neglecting terms of higher order than $O(\varepsilon^2)$ these equations become:

$$\frac{\partial}{\partial \zeta} r_{upp} = r_{upp} \left(2 i \kappa_a + 2A\right) + B r_{usp} + E r_{ups} + C,$$

$$\frac{\partial}{\partial \zeta} r_{ups} = r_{ups} \left(i \kappa_a + i \kappa_b + A + F\right) + B r_{uss} + r_{upp} B + D,$$

$$\frac{\partial}{\partial \zeta} r_{usp} = r_{usp} \left(i \kappa_b + i \kappa_a + F + A\right) + E r_{upp} + r_{uss} E + G,$$

$$\frac{\partial}{\partial \zeta} r_{uss} = r_{uss} \left(2 i \kappa_b + 2F\right) + B r_{usp} + E r_{ups} + H,$$

$$\frac{\partial}{\partial \zeta} \Psi_{dp} = r_{upp} C + r_{ups} G + A + t_{dsp} e^{-i \kappa_a \zeta} B,$$ (5.27)

$$\frac{\partial}{\partial \zeta} t_{dps} = t_{dps} \left(i \kappa_a + A\right) + e^{i \kappa_b \zeta} \left(r_{upp} D + r_{ups} H + B + B \Psi_{ds}\right),$$

$$\frac{\partial}{\partial \zeta} t_{dsp} = t_{dsp} \left(i \kappa_b + F\right) + e^{i \kappa_a \zeta} \left(E + E \Psi_{dp} + r_{usp} C + r_{uss} G\right),$$

$$\frac{\partial}{\partial \zeta} \Psi_{ds} = r_{usp} D + r_{uss} H + F + t_{dps} e^{-i \kappa_b \zeta} E.$$

It should be noted again that the quantities A, B, C, D, E, F, G and H are combinations of terms of first and second powers of the medium fluctuations. Therefore, in the system (5.27) not all terms of the third order in ε have been neglected. However, as it is very convenient to perform the derivation in terms of the functions A, B, C, D, E, F, G and H, we can still neglect the remaining high-order terms in the final equations.

In order to derive the reflectivities of the P- and SV-waves, caused by an incident plane wave in the uppermost half space, we need also the other equations from system (5.23). We consider these reflectivities in Chapter 10.

Taking initial conditions (5.25) into account and looking upon the equations in system (5.27) as 8 independent equations, we find formal first-order solutions for the quantities $r_{upp}, r_{ups}, t_{dsp} \ r_{uss}, r_{usp}$ and t_{dps}. Then, we substitute them into the expressions for Ψ_{dp} and Ψ_{ds}. After expanding the exponential terms in the Taylor series and neglecting terms of higher order than $O(\varepsilon^2)$ we obtain the final results for Ψ_{dp} and Ψ_{ds}:

$$
\begin{aligned}
\Psi_{dp}(L) = \int_0^L d\zeta \{ A(\zeta) + \int_0^\zeta d\zeta_1 [C'(\zeta) C'(\zeta_1) \exp(-2i\kappa_a(\zeta_1 - \zeta)) \\
+ P'D'(\zeta)D'(\zeta_1) \exp(-i(\kappa_a + \kappa_b)(\zeta_1 - \zeta)) \\
+ P'B'(\zeta)B'(\zeta_1) \exp(-i(\kappa_b - \kappa_a)(\zeta_1 - \zeta))]\},
\end{aligned}
$$

$$
\begin{aligned}
\Psi_{ds}(L) = \int_0^L d\zeta \{ F(\zeta) + \int_0^\zeta d\zeta_1 [H'(\zeta) H'(\zeta_1) \exp(-2i\kappa_b(\zeta_1 - \zeta)) \\
+ P'D'(\zeta)D'(\zeta_1) \exp(-i(\kappa_a + \kappa_b)(\zeta_1 - \zeta)) \\
+ P'B'(\zeta)B'(\zeta_1) \exp(-i(\kappa_a - \kappa_b)(\zeta_1 - \zeta))]\},
\end{aligned}
$$

(5.28)

where the quantities B', C', D' and H', are first-order approximations of the quantities B, C, D and H, respectively:

$$
B' = -\frac{ipb\omega}{2Xa}(C_4\varepsilon_b + C_3\varepsilon_\varrho),
$$

$$
C' = -\frac{i\omega}{Xa}(\varepsilon_a + C_2\varepsilon_b + C_1\varepsilon_\varrho),
$$

$$
D' = -\frac{ipb\omega}{2Xa}(C_6\varepsilon_b + C_5\varepsilon_\varrho),
$$

$$
H' = \frac{i\omega}{bY}(C_8\varepsilon_b + C_7\varepsilon_\varrho),
$$

$$
P' = \frac{Xa}{bY},
$$

(5.29)

and the following abbreviations are used

$$
\begin{aligned}
X &= \sqrt{1 - p^2a^2}, \quad Y = \sqrt{1 - p^2b^2}, \\
C_1 &= X^2(1 - 4p^2b^2), \quad C_2 = -8p^2b^2X^2, \\
C_3 &= X(3 - 4b^2p^2) - Y(a/b + 2b/a - 4abp^2), \\
C_4 &= 4X(1 - 2b^2p^2) - 4Y(b/a - 2abp^2), \\
C_5 &= X(4b^2p^2 - 3) - Y(a/b + 2b/a - 4abp^2), \\
C_6 &= 4X(2b^2p^2 - 1) - 4Y(b/a - 2abp^2), \\
C_7 &= Y^2(1 - 4p^2b^2), \quad C_8 = 1 - 8p^2b^2Y^2.
\end{aligned}
$$

(5.30)

Therefore, the transmitted wavefields $T_{p,s}$, caused by the P- and SV-waves incident from the uppermost half space, are

$$T_{p,s}(x, L) = \exp\left(i\kappa_{a,b}L + \Psi_{dp,ds}(L) + i\omega(px - t)\right), \qquad (5.31)$$

where the phase-correction and attenuation terms are described by the imaginary and real parts of the quantities $\Psi_{dp}(L)$ and $\Psi_{ds}(L)$ given by formulas (5.28). In the following we consider a real-valued $\kappa_{a,b}$ only. Then the phase velocities and the attenuation coefficients are

$$c_{p,s} = 1/\sqrt{p^2 + \frac{Im^2\{i\kappa_{a,b}L + \Psi_{dp,ds}(L)\}}{L^2\omega^2}}, \quad \gamma_{p,s} = Re\{\Psi_{dp,ds}(L)\}/L. \qquad (5.32)$$

From now on we can already consider the quantities $c_{p,s}$, and $\gamma_{p,s}$ as phase velocities and attenuation coefficients of the transmitted wavefields in the *inhomogeneous dynamic-equivalent medium* replacing the inhomogeneous stratification. The complex-valued vector

$$\mathbf{k}_{p,s} = (\omega p, \kappa_{a,b} - i\Psi_{dp,ds}) \qquad (5.33)$$

is then the wave vector in such a medium (from now on we consider real-valued $\kappa_{a,b}$ only). Generally, however, this wave vector, as well as the phase velocities and the attenuation coefficients, depends on L. Therefore we have to define this medium as a travel-distance-dependent (i.e., inhomogeneous) dynamic-equivalent medium. In the case of random stationary fluctuations the statistical expectation of the vector $\mathbf{k}_{p,s}$ is independent of L and the dynamic-equivalent medium can be defined as a homogeneous one (see the next section).

The real and imaginary parts of the quantity Ψ_{dp} given by equations (5.28) are coupled by the Hilbert transform:

$$\frac{1}{\omega}\left(\frac{Im\{\Psi_{dp}(\omega)\}}{\omega} - \frac{\partial Im\{\Psi_{dp}(\omega)\}}{\partial\omega}\Big|_{\omega=0}\right) = \frac{1}{\pi}P.V.\int_{-\infty}^{\infty}\frac{Re\{\Psi_{dp}(\Omega)\}}{\Omega^2(\Omega - \omega)}d\Omega. \qquad (5.34)$$

This formula is a kind of *Kramers-Krönig dispersion relation* (see e.g., Aki and Richards, 1980, pp.173-175). Taking into account that the same relationship is valid for the real and imaginary parts of the components of the wave vector $\mathbf{k}_p$ we conclude that the transmissivity T_p is not only causal but it is also *minimum phase* (at least in our approximation). It is in agreement with the theorem that for propagation normal to the layer boundaries the transmissivity is minimum phase (see e.g., Robinson and Treitel 1980, pp.321-325). Generally this is, however, not the case for the wavefield T_s (note the difference in the exponential factors of terms with the function B' in expressions (5.28) for Ψ_{dp} and Ψ_{ds} respectively). It is also interesting to note here that sometimes the Kramers-Krönig relationship can be used to provide a dynamic-equivalent-medium description (see e.g., Beltzer, 1989). In such cases the minimum-phase property of transmissivities must be assumed.

The validity of formulas (5.28)-(5.31) for the transmissivities is not restricted by the assumption of stationarity of the medium fluctuations. Moreover, all formulas can be applied to calculate the wavefields in deterministic models like, e.g., in a medium with a linear gradient.

The invariant-embedding method, combined with the small perturbation expansion, allowed us to obtain the above results for the transmissivity. As previously noted, the invariant-embedding equations (5.23) are rather general. They permit to derive different types of approximations of wavefields, e.g., low-frequency as well as small-perturbation ones. If we start from an earlier point of derivations with a small-perturbation assumption then some results can be obtained in a more direct way. In Appendix 5.9 we show such a relatively straightforward derivation of results (5.28)-(5.31).

5.5   Transmissivity in Stationary Random Media

Assuming now stationarity of the medium fluctuations, taking limits (5.52) defined below in Appendix 5.9 and applying the statistical averaging to equation (5.28), we obtain the following time-harmonic transmissivity

$$T_{p,s}(x, L) = \exp\left[i\psi_{p,s}L + i\omega(px - t) - \gamma_{p,s}L\right], \tag{5.35}$$

where the quantities $\psi_{p,s}$ and $\gamma_{p,s}$ are the angle- and frequency-dependent vertical-phase increments and attenuation coefficients describing the wave propagation in the homogeneous dynamic-equivalent medium, which replaces the inhomogeneous laminations. For the P-wave these quantities are

$$\psi_p = \kappa_a + \omega A_P -$$
$$- \omega^2 \int_0^\infty d\xi \left[B_P(\xi)\sin(2\xi\kappa_a) + B_{BB}(\xi)\sin(\xi\kappa_-) + B_{DD}(\xi)\sin(\xi\kappa_+)\right] \tag{5.36}$$
$$\gamma_p = \omega^2 \int_0^\infty d\xi \left[B_P(\xi)\cos(2\xi\kappa_a) + B_{BB}(\xi)\cos(\xi\kappa_-) + B_{DD}(\xi)\cos(\xi\kappa_+)\right],$$

and for the SV-wave they are:

$$\psi_s = \kappa_b + \omega A_{SV} -$$
$$- \omega^2 \int_0^\infty d\xi \, [B_{SV}(\xi)\sin(2\xi\kappa_b) - B_{BB}(\xi)\sin(\xi\kappa_-) + B_{DD}(\xi)\sin(\xi\kappa_+)]$$
$$\tag{5.37}$$
$$\gamma_s = \omega^2 \int_0^\infty d\xi \, [B_{SV}(\xi)\cos(2\xi\kappa_b) + B_{BB}(\xi)\cos(\xi\kappa_-) + B_{DD}(\xi)\cos(\xi\kappa_+)],$$

where $\kappa_a = \omega\sqrt{(1/a^2 - p^2)}$, $\kappa_b = \omega\sqrt{(1/b^2 - p^2)}$, $\kappa_+ = \kappa_b + \kappa_a$ and $\kappa_- = \kappa_b - \kappa_a$ (we consider here real-valued $\kappa_{a,b}$ only). The quantities B_P, B_{BB}, B_{DD} and B_{SV} are the following combinations of auto- and crosscorrelation functions of the medium fluctuations:

$$B_P(\xi) = \frac{1}{X^2 a^2}\left[B_{\varrho\varrho}(\xi)C_1^2 + 2B_{a\varrho}(\xi)C_1 + 2B_{b\varrho}(\xi)C_1 C_2 \right.$$
$$\left. + 2B_{ab}(\xi)C_2 + B_{aa}(\xi) + B_{bb}(\xi)C_2^2 \right],$$

$$B_{SV}(\xi) = \frac{1}{Y^2 b^2}\left[B_{\varrho\varrho}(\xi)C_7^2 + 2B_{b\varrho}(\xi)C_7 C_8 + B_{bb}(\xi)C_8^2 \right], \tag{5.38}$$

$$B_{BB}(\xi) = \frac{p^2 b}{4XYa}\left[B_{\varrho\varrho}(\xi)C_3^2 + 2B_{b\varrho}(\xi)C_3 C_4 + B_{bb}(\xi)C_4^2 \right],$$

$$B_{DD}(\xi) = \frac{p^2 b}{4XYa}\left[B_{\varrho\varrho}(\xi)C_5^2 + 2B_{b\varrho}(\xi)C_5 C_6 + B_{bb}(\xi)C_6^2 \right].$$

The functions $B_{aa}, B_{bb}, B_{ab}, B_{a\varrho}, B_{\varrho\varrho}$, and $B_{b\varrho}$ are the auto- and crosscorrelation functions of the fluctuations:

$$B_{aa}(\xi) = \langle \varepsilon_a(z)\varepsilon_a(z+\xi) \rangle,$$
$$B_{bb}(\xi) = \langle \varepsilon_b(z)\varepsilon_b(z+\xi) \rangle,$$
$$B_{\varrho\varrho}(\xi) = \langle \varepsilon_\varrho(z)\varepsilon_\varrho(z+\xi) \rangle,$$
$$B_{ab}(\xi) = \langle [\varepsilon_a(z)\varepsilon_b(z+\xi) + \varepsilon_a(z+\xi)\varepsilon_b(z)]/2 \rangle, \tag{5.39}$$
$$B_{a\varrho}(\xi) = \langle [\varepsilon_a(z)\varepsilon_\varrho(z+\xi) + \varepsilon_a(z+\xi)\varepsilon_\varrho(z)]/2 \rangle,$$
$$B_{b\varrho}(\xi) = \langle [\varepsilon_b(z)\varepsilon_\varrho(z+\xi) + \varepsilon_b(z+\xi)\varepsilon_\varrho(z)]/2 \rangle$$

For practical needs the angular brackets (denoting averaging over a statistical ensemble) are replaced by spatial averaging thus providing the usual auto- and crosscorrelation functions. These can be estimated from the fluctuations of the one and only set of the sonic-, shear- and density logs. The quantities A_P and

A_{SV} are the following combinations of the variances and crossvariances of the medium fluctuations

$$A_P = \frac{1}{2Xa}[\sigma_{\varrho\varrho}^2(1 - 4p^2b^2Z) + 2\sigma_{a\varrho}^2(1 - 4p^2b^2) + 8\sigma_{b\varrho}^2p^2b^2(1 - 2Z) +$$
$$3\sigma_{aa}^2 - 16p^2b^2\sigma_{ab}^2 + 16p^2b^2\sigma_{bb}^2(1 - Z)], \qquad (5.40)$$
$$A_{SV} = \frac{1}{2Yb}\left[\sigma_{\varrho\varrho}^2(1 - C_9) + \sigma_{b\varrho}^2(2 - 4C_9) + \sigma_{bb}^2(3 - 4C_9)\right],$$

where the quantities X, Y, C_1-C_8 were defined in equations (5.30), and

$$Z = p^2(a^2 - b^2), \quad C_9 = 4Y^2Zb^2/a^2. \qquad (5.41)$$

Here the variances and crossvariances themselves are defined by $\sigma_{aa}^2 = B_{aa}(0)$, $\sigma_{bb}^2 = B_{bb}(0)$, $\sigma_{\varrho\varrho}^2 = B_{\varrho\varrho}(0)$, $\sigma_{ab}^2 = B_{ab}(0)$, $\sigma_{a\varrho}^2 = B_{a\varrho}(0)$ and $\sigma_{b\varrho}^2 = B_{b\varrho}(0)$. It should be noted that the crossvariances $\sigma_{\varrho\varrho}^2$, σ_{ab}^2, and $\sigma_{a\varrho}^2$ can be also negative.

Vertical incidence In the case of vertical incidence the horizontal slowness p vanishes. Thus, the quantities B_{BB}, B_{DD}, Z, C_2 and C_9 are equal to zero and C_1, C_7, C_8, X and Y are equal to one. Substituting this into formulas (5.36) and (5.37), we immediately arrive at the expressions $\psi_{ac,SH}$ and $\gamma_{ac,SH}$ for vertically-incident acoustic pressure- and elastic SH-waves (see equations (4.18)-(4.20)).

Constant density and shear velocity In the case of oblique incidence in a medium with constant density and shear velocity the auto- and crosscorrelation functions $B_{\varrho\varrho}, B_{bb}, B_{b\varrho}$, along with the corresponding variances and covariances $\sigma_{\varrho\varrho}^2, \sigma_{bb}^2$ and $\sigma_{b\varrho}^2$ are equal to zero. All four functions B_{BB}, B_{DD}, B_{SV} and A_{SV} are therefore equal to zero too. As in the previously considered situation the wave propagation is characterized by the absence of P- and SV- wave coupling. The SV-wave propagates as in the homogeneous reference medium. Results (5.36) correspond to those for an acoustic medium (see equations (4.18) and (4.20)) with a constant density.

General reduction to the acoustic case Let the shear modulus tend to zero. This causes the shear velocity and correlation functions $B_{bb}, B_{b\varrho}$, along with the variances $\sigma_{bb}^2, \sigma_{b\varrho}^2$ and the quantities C_2, C_4, C_6 to become zero. The following simplifications result:

$$B_P = \frac{1}{X^2 a^2} \left[B_{\varrho\varrho} X^4 + 2B_{a\varrho} X^2 + B_{aa} \right],$$

$$B_{BB} = \frac{p^2 b}{4XYa} B_{\varrho\varrho}, \tag{5.42}$$

$$B_{DD} = B_{BB}.$$

We now have to take the limit in formulas (5.36). For this we have to take into account that λ_+ and λ_- tend to infinity like λ_b (i.e., like $1/b$). The asymptotic estimates of the corresponding Fourier sine transforms of the functions $B_{BB}(\xi)$ and $B_{DD}(\xi)$ are $B_{BB}(0)/\lambda_b$ and $B_{DD}(0)/\lambda_b$, respectively. The Fourier cosine transforms of the functions $B_{BB}(\xi)$ and $B_{DD}(\xi)$ tend to zero like $1/\lambda_b^2$ or faster. Using this information and relations (5.42) we immediately obtain results (4.18).

Therefore, in the cases of vertical incidence, of constant density and shear velocity or in the limit of a zero shear modulus the results (5.36) and (5.37) are reduced to a scalar-wave transmissivity. Moreover, considering the above results for the SV-wave and the results for the SH-wave from the previous chapter, the need arises to accept a *frequency-dependent shear-wave splitting* for the considered media. This topic is further discussed below in Section 6.4

5.6   Transmissivity for Non-Stationary Random Media

Besides imperfect parallel layering (which is neglected in this book) real laminations are in generally statistically inhomogeneous. This implies that the statistical moments of the density-, compressional and shear-velocity fluctuations are depth-dependent. The general solution for the transmissivity in this case is given by equations (5.28)-(5.31). In the following we consider the attenuation of the transmissivity for random media for which the ensemble-averaged densities and velocities linearly increase with depth.

In equations (5.28) the first terms, containing the quantities $A(\zeta)$ and $F(\zeta)$ respectively, are purely imaginary and yield phase shifts only. The next integral terms have similar structures and they can be analyzed simultaneously. Let us consider the first one:

$$\int_0^L d\zeta \int_0^\zeta d\zeta_1 C'(\zeta) C'(\zeta_1) \exp(-2i\kappa_a(\zeta_1 - \zeta)). \tag{5.43}$$

The quantity $C'(\zeta)$ is a combination of the first-order powers of the medium fluctuations. It can be written in the following form:

$$C'(\zeta) = \varepsilon_z(\zeta - L/2) + \varepsilon_c(\zeta), \tag{5.44}$$

where the quantity ε_z represents the gradient (for the sake of simplicity we assume that both the density and the velocities have the same gradient) and the quantity $\varepsilon_c(\zeta)$ represents random fluctuations of the quantity C'. By definition, ε_c must be a realization of a stationary random process with $\langle \varepsilon_c \rangle = 0$. Both the quantities $\varepsilon_z L/2$ and ε_c are of the order $O(\varepsilon)$.

Substituting now equation (5.44) into expression (5.43), neglecting all terms of the orders higher than $O(\varepsilon^2)$ and performing the statistical averaging, we arrive at an expression which contains two terms. The first term is a combination of auto- and crosscorrelation functions of random fluctuations of the densities and the velocities given by the autocorrelation of ε_c. This term is already well known for stationary media (its real and imaginary parts are equal to the first integral terms of equations (5.36) multiplied by L). The second term has the form

$$i\kappa_a L^3 \frac{\varepsilon_z^2}{24} + O(L^0). \tag{5.45}$$

The leading term of this sum is purely imaginary and does not influence the attenuation. Other contributions in this expression are much smaller than the above mentioned stationary term, which is of the order $O(L)$. Therefore, we can immediately conclude that the indicated type of nonstationarity does not influence the attenuation. However, it needs to be taken into account for calculations of the phase shift. This conclusion was already arrived at by Banik et al. (1985) and Burridge et al. (1988) for the case of a vertically propagating plane pressure wave in a non-stationary randomly layered fluid. Here we have arrived at the same conclusion for obliquely propagating elastic waves.

To illustrate this we compare now the attenuation coefficient of a numerically layer-code-computed P-wave transmissivity for a *non-stationary medium* with the attenuation coefficient analytically calculated by the generalized ODA formula (5.36) for a *purely stationary medium* with the same statistics for the small-scale fluctuations.

The depth dependency of the compressional velocity in Figure 9 is the same as in Figure 10, but with a superimposed linear gradient. The density and the shear velocity have identical (but scaled) depth-dependencies. The model parameters are: $a = 4000m/s, b = 2300m/s, \varrho = 2.5g/cm^3, l = 10m, L = 500m, \varepsilon_z L = 0.6$.

The standard deviations of the velocities and density are equal to 15%. Figure 11 shows analytical and numerical results for the frequency-dependent attenuation coefficient for $\vartheta = 30°$. One can observe that the (smooth) theoretical result predicts very well the smoothed behavior of the numerically computed one.

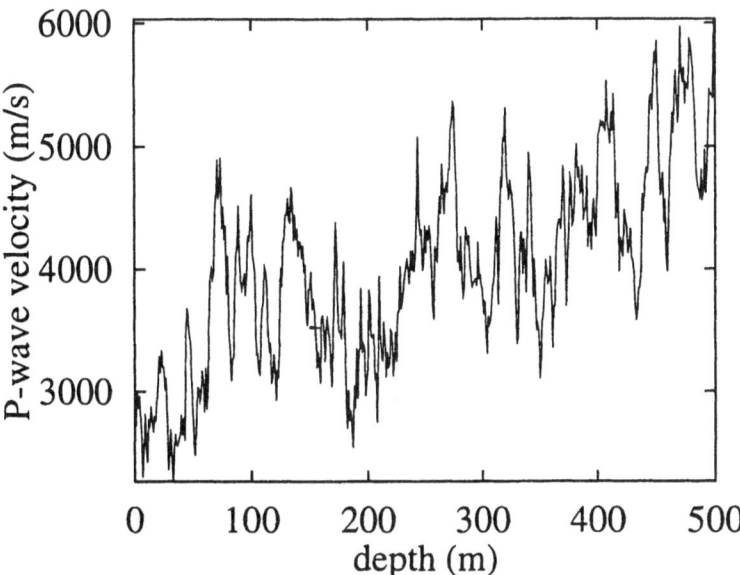

Fig. 9. Non-stationary synthetic sonic log. A linear positive gradient is superimposed with a realization of a random stationary process with an exponential correlation function. The averaged compressional velocity is $4000m/s$, the correlation distance is 10m and the standard deviation is 15 percent. (Reprinted from Shapiro et al., 1996, with the kind permission of Blackwell Science.)

It can be expected that in other cases of nonstationarity the attenuation is also only caused by the random small-scale (much smaller than the travel distance) stationary fluctuations of the density and the velocities superimposed on smooth (on a large scale, i.e., the scale of the order of the traveldistance) depth-dependent functions. This means that in order to correct the stratigraphic-filtering effect a combination of ray tracing and the generalized O'Doherty-Anstey formula (for stationary media) should be considered. *In this way we can look upon the above results as implying a generalization of standard ray theory to 1-D random non-stationary elastic media.* This should be stressed as it is a common belief that ray theory is not valid in stratified media. In fact standard ray solutions, valid for concentrated sources and smooth 1-D medium parameters

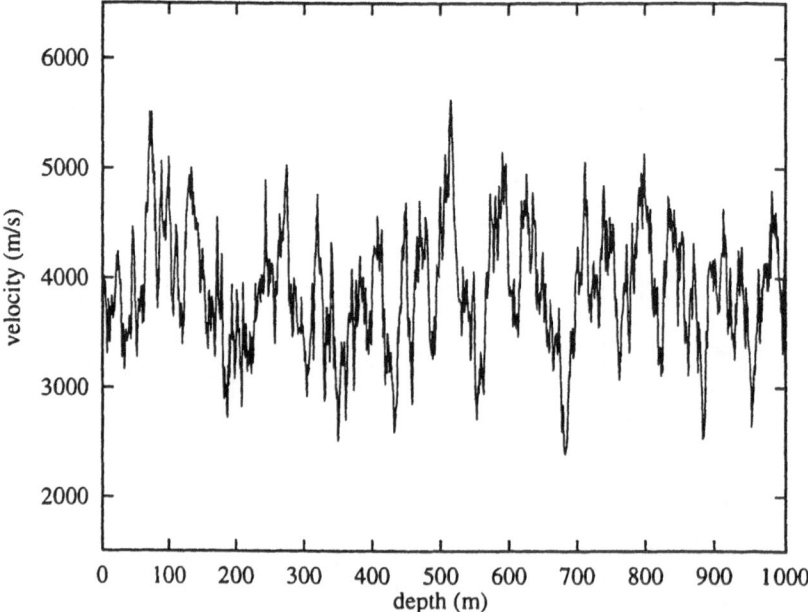

Fig. 10. Synthetic sonic log as in Figure 9 but now without the linear positive gradient. (Reprinted from Shapiro et al., 1996, with the kind permission of Blackwell Science.)

can easily be modified to take small-scale stationary fluctuations into account.

In order to do this, it is clear from the above consideration that a statistical description of small-scale fluctuations is needed. Such a description includes first and second statistical moments of density and velocities fluctuations. We call such a description a *statistical macro model*. In Chapter 8 we apply this approach in the case of velocity- and density-depth dependencies observed in one of regions of the North Sea.

5.7   Validity Conditions of the Solution

The solutions given in this chapter have a restricted validity range. The most important point here is the expansion of the exponential factors in equations (5.23) in order to obtain equations (5.28) and later equations (5.36) and (5.37). This expansion is in conflict with the limits in equation (5.52). This implies that our solutions are valid only in an intermediate asymptotic range confining L to

$$\max(\lambda, l) < L < \max(\lambda, l)/\varepsilon^2. \tag{5.46}$$

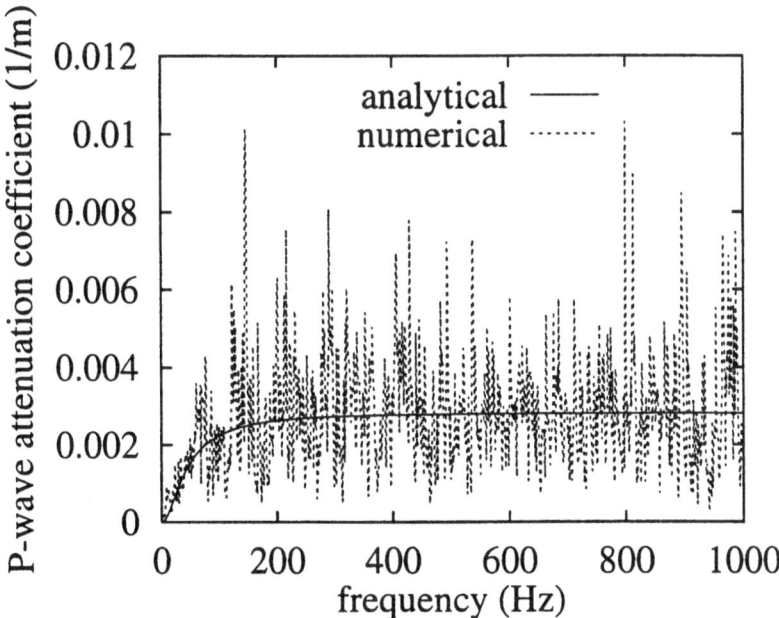

Fig. 11. Comparison of the numerically and theoretically computed frequency-dependent P-wave attenuation coefficient for $\vartheta = 30°$. For the numerical curve the medium parameters in Figure 9 were used. For the theoretical curve the statistical parameters of the corresponding stationary process (Figure 10) have been used. (Reprinted from Shapiro et al., 1996, with the kind permission of Blackwell Science.)

The left-hand boundary of this inequality is necessary in order to evaluate the limits in equations (5.52). Thus, it is important for equations (5.36) and (5.37), which represent the statistical version of the generalized ODA formulas. The right-hand boundary on the other hand is necessary to keep the expansion valid for the exponential terms in equations (5.23). It is also important for the deterministic version of the approximation, i.e., equations (5.28). However, note that the obtained solutions (5.28), (5.36) and (5.37) are accurate up to the second order in ε. In general, an extension of the validity range for L would require taking into account terms smaller than $O(\varepsilon^2)$. It can also be observed that the validity range of L is automatically extended for the cases where the problem is reduced to low frequencies. It seems to be that this is also the case for $b \to 0$, i.e., for the case, where the problem is reduced to scalar waves. Moreover, it is important to understand that the attenuation coefficients given by the ODA-approximation can differ significantly from those describing strongly localized wavefields.

At this point we want to recall all other restrictions of our solutions. These are: ideally elastic media (see, however, Section 8.6 and Chapter 9), plane-wave transmissivity, undercritical incidence (for SV-waves; however, the theory can be relatively easy extended to the case of overcritical SV-waves), small fluctuations of the density, compressional- and shear-wave velocities. We also assumed additionally that the correlation functions of the medium fluctuations are quickly vanishing functions of the correlation lag. However, in spite of all above restrictions the proposed analytic solutions are robust and describe well those transmissivities that are relevant for seismology and exploration seismics. This will be illustrated in the following chapters.

5.8   Appendix: Explicit Expressions for Some Quantities.

After neglecting terms with higher than second-order powers of fluctuations (see equation 5.7), the elements of the matrix $\mathbf{Q}_\epsilon$ take the following forms:

$$\varepsilon_{13} = \left[3\varepsilon_b^2 + \varepsilon_\varrho^2 + 2\varepsilon_\varrho\varepsilon_b - 2\varepsilon_b - \varepsilon_\varrho\right]/(\varrho b^2),$$

$$\varepsilon_{21} = 2\omega p b^2 \left[3\varepsilon_a^2 + \varepsilon_b^2 - 4\varepsilon_a\varepsilon_b + 2\varepsilon_b - 2\varepsilon_a\right]/a^2,$$

$$\varepsilon_{24} = \left[3\varepsilon_a^2 + \varepsilon_\varrho^2 + 2\varepsilon_\varrho\varepsilon_a - 2\varepsilon_a - \varepsilon_\varrho\right]/(\varrho a^2),$$

$$\varepsilon_{31} = [8p^2b^4\left(4\varepsilon_b + \varepsilon_\varrho + 1\right)\varepsilon_a - 12p^2b^4\varepsilon_a^2 + 4p^2b^2\left(a^2 - 6b^2\right)\varepsilon_b^2$$
$$\quad + 8p^2b^2\left(a^2 - 2b^2\right)\left(\varepsilon_\varrho + 1\right)\varepsilon_b + \left(4p^2b^2a^2 - a^2 - 4p^2b^4\right)\varepsilon_\varrho]\varrho\omega^2/a^2,$$

$$\varepsilon_{42} = -\varrho\omega^2\varepsilon_\varrho.$$

Below we give the expressions for the elements of the matrix $\mathbf{\Lambda}_\epsilon(z)$, (equation

5.12) neglecting terms with higher than second-order powers of fluctuations:

$$A = -[2\left(8p^2b^2\varepsilon_b + 4\varepsilon_\varrho b^2 p^2 - \varepsilon_\varrho + 1\right)\varepsilon_a - 3\varepsilon_a^2 + 16b^2p^2\left(Z - 1\right)\varepsilon_b^2$$
$$+ 8b^2p^2(2Z - 1)\varepsilon_\varrho\varepsilon_b + (4p^2b^2Z - 1))\varepsilon_\varrho^2]i\omega/(2Xa),$$

$$B = -[4b^2Y\left(2\varepsilon_b + \varepsilon_\varrho\right)\varepsilon_a + 2b\left(-3aXM + bY(2p^2a^2 - 1 - 8p^2b^2)\right)\varepsilon_b^2$$
$$+ b\left(4\left(-aXM + 2p^2bY(a^2 - 2b^2)\right)\varepsilon_\varrho + aC_4\right)\varepsilon_b$$
$$+ 2bM(-aX + bY)\varepsilon_\varrho^2 + abC_3\varepsilon_\varrho]ip\omega/(2Xa^2),$$

$$C = -[2\left(8p^2b^2\varepsilon_b + (4b^2p^2 - 1)\varepsilon_\varrho + 1\right)\varepsilon_a - 3\varepsilon_a^2 + 8b^2p^2\left(M - p^2a^2\right)\varepsilon_b^2$$
$$+ 8b^2p^2\left(M\varepsilon_\varrho - 2X^2\right)\varepsilon_b + \left(4b^2p^2(X^2 + Y^2) - 1\right)\varepsilon_\varrho^2$$
$$+ 2C_1\varepsilon_\varrho]i\omega/(2aX),$$

$$\text{(5.47)}$$

$$D = -[4Yb^2\left(2\varepsilon_b + \varepsilon_\varrho\right)\varepsilon_a + 2b\left(3aXM + bY(2p^2a^2 - 1 - 8p^2b^2)\right)\varepsilon_b^2$$
$$+ b\left(4\left(aXM + 2\bar{p}^2bY(a^2 - 2b^2)\right)\varepsilon_\varrho + aC_6\right)\varepsilon_b$$
$$+ 2bM(aX + b)\varepsilon_\varrho^2 + abC_5\varepsilon_\varrho]ip\omega/(2Xa^2),$$

$$F = -[a^2\left(4C_9 - 3\right)\varepsilon_b^2 + \left(a^2(4C_9 - 2)\varepsilon_\varrho + 2a^2\right)\varepsilon_b$$
$$+ (C_9 - 1)\varepsilon_\varrho^2]i\omega/(2ba^2Y),$$

$$H = -[(-8p^2b^4(M + X^2) - 8p^2b^2a^2 + 3a^2)\varepsilon_b^2$$
$$+ 2\left((a^2 - 8p^2b^4Y^2)\varepsilon_\varrho - a^2C_8\right)\varepsilon_b$$
$$- \left(4p^2b^4(1 - X^2 - Y^2) - 4p^2b^2a^2 + a^2\right)\varepsilon_\varrho^2 - 2a^2C_7\varepsilon_\varrho]i\omega/(2ba^2Y),$$

where $M = 1 - 2b^2p^2$. Furthermore, $E = XaB/(bY)$ and $G = XaD/(bY)$. All other notations are given in the definitions (5.30) and (5.41).

5.9 Appendix: Alternative Derivation of Equations (5.35)-(5.37)

In this chapter we have shown that the transmissivity given by equations (5.35)-(5.37) can be obtained by applying a combination of the invariant-embedding method and the small-perturbation approximation to equation (5.1). Here we show that another even more straightforward approach is possible.

Let us recall our basic assumptions: the medium fluctuations are small and the incidence of SV-wave is undercritical. This implies that the quantities $A(z)$, $B(z)$, $C(z)$, $D(z)$, $E(z)$, $F(z)$, $G(z)$, and $H(z)$ are small (of the order $O(\varepsilon)$) too. Then we can solve the system (5.10) by the small-perturbation approximation directly. In the following we obtain the attenuation coefficients and vertical phase increments of the P- and SV-wave transmissivities which include terms of the order not higher than $O(\varepsilon^2)$. This leads again to the generalized ODA formula. We stress that we do not consider the P-SV and SV-P transmissivities. However, our results for the P-P and SV-SV transmissivities take into account the P-SV and SV-P conversions due to internal scattering in the inhomogeneous medium.

First, we derive the transmissivity of the P-wave. For this we consider the following boundary conditions (note that the vector $\mathbf{g}(z) = (d_p(z), d_s(z), u_p(z), u_s(z))^T$ is continuous at any interface in the medium):

$$d_p(0) = 1, \quad d_s(0) = 0, \quad u_p(L) = 0, \quad u_s(L) = 0. \tag{5.48}$$

Therefore, at $z = L$ the radiation conditions of vanishing upgoing P- and SV-wavefields are satisfied. At $z = 0$ the single down-going incident P-wave is given. For such boundary conditions only this mode, $d_p(z)$, differs from zero in the case of vanishing fluctuations. However, in inhomogeneous media the quantities $d_s(z)$, $u_p(z)$ and $u_s(z)$ cannot be equal to zero due to the internal P-P, P-SV, SV-SV and SV-P scattering processes, which take place at any depth in the interval $0 < z < L$. These quantities are of the order of the fluctuations, $O(\varepsilon)$.

Taking into account equation (5.48), neglecting terms of higher order than $O(\varepsilon)$ and integrating equation (5.10), we obtain the first-order approximations of the down and upgoing waves:

$$d_p^I(z) = e^{i\kappa_a z + \int_0^z A(z')dz'},$$

$$d_s^I(z) = e^{i\kappa_b z} \int_0^z E(z')e^{i(\kappa_a - \kappa_b)z'}dz',$$

$$u_p^I(z) = e^{-i\kappa_a z} \int_z^L C(z')e^{2i\kappa_a z'}dz', \tag{5.49}$$

$$u_s^I(z) = e^{-i\kappa_b z} \int_z^L G(z')e^{i(\kappa_a + \kappa_b)z'}dz'.$$

It should be noted here that the quantities A, B, C, D, E, F, G and H are combinations of terms with the first and second powers of the medium fluctuations. Therefore, in the system of equations (5.49) not all terms of the second order in ε have been neglected. However, as it is very convenient to perform the derivation in terms of the functions A, B, C, D, E, F, G and H, we can always neglect the remaining high-order terms in the final equations.

Further, to consider the transmissivity of the P-wave we need the second-order approximation of $d_p(z)$ only. Moreover, in order to find the (dynamic equivalent-medium) angle-dependent phase velocity and attenuation coefficient of the transmissivity, we have to analyze its logarithm. For this it is convenient to rewrite the first equation of system (5.10) in the following form:

$$\frac{\partial(r_p(z) + i\phi_p(z))}{\partial z} = i\kappa_a + A(z) + C(z)\frac{u_p(z)}{d_p(z)} + B(z)\frac{d_s(z)}{d_p(z)} + D(z)\frac{u_s(z)}{d_p(z)}, \quad (5.50)$$

where $r_p(z)$ is the real part of $\ln d_p(z)$ and $\phi_p(z)$ is its imaginary part. By substituting in this equation the first-order approximation (5.49) of the quantities $d_p(z)$, $d_s(z)$, $u_p(z)$ and $u_s(z)$ and neglecting terms of higher order than $O(\varepsilon^2)$ we obtain:

$$
\begin{aligned}
r_p(L) + i\phi_p(L) = \quad & i\kappa_a L + \int_0^L dz[A(z) \\
& + B(z)e^{i(\kappa_b - \kappa_a)z}\int_0^z E(z')e^{i(\kappa_a - \kappa_b)z'}dz' \\
& + C(z)e^{-2i\kappa_a z}\int_z^L C(z')e^{2i\kappa_a z'}dz' \\
& + D(z)e^{-i(\kappa_a + \kappa_b)z}\int_z^L G(z')e^{i(\kappa_a + \kappa_b)z'}dz'].
\end{aligned}
\quad (5.51)
$$

This equation provides an approximation for the transmissivity in deterministic as well as random (also non-stationary) media.

We are now interested in the solution for the attenuation coefficients and the phase increments for thick stationary random media. This means that the subject of our study is now the evaluation of the limits

$$\psi_p = \lim_{L\to\infty} \frac{\phi_p(L)}{L}, \quad \gamma_p = -\lim_{L\to\infty} \frac{r_p(L)}{L}. \quad (5.52)$$

Therefore, we insert approximation (5.51) into equations (5.52). Instead of providing here the detailed chain of the simplifications of equations (5.52) we point out the principal steps (a similar derivation was shown in Appendix 4.6).

(1) Because of the self–averaging property the limits (5.52) are equal to their statistically averaged values. This consequently allows us to use ensemble averaging for an additional simplification of equations (5.52) in order to express the

attenuation coefficients and phase increments in terms of a restricted number of statistical parameters of the medium fluctuations (i.e. to obtain the smoothed attenuation coefficients and phase increments). After this averaging we obtain instead of the random functions $A(z), B(z), C(z), D(z), E(z), F(z), G(z)$ and $H(z)$ their mathematical expectations. In the same way we obtain instead of the product of any two of these functions their autocorrelation or crosscorrelation functions. Taking into account that the medium fluctuations (and therefore $A(z), B(z), C(z), D(z), E(z), F(z), G(z)$ and $H(z)$) are stationary functions of depth, it is clear that their mathematical expectations are independent of the depth and their correlation functions depend on the depth increment ξ between any two depth levels z and $z' = z + \xi$.

(2) The two following properties of the correlation functions are important to simplify (5.52). The correlation functions are even functions of the argument and their absolute values quickly vanish outside a finite domain of the argument. Certain media (e.g., periodic media) may violate this last property. However, for realistic stratifications of sediments our considerations are valid.

After performing the indicated calculations the estimation of the limits (5.52) yields equations (5.37). The physical meaning of the quantities ψ_p and γ_p is clear from the equations (5.52): ψ_p is the dynamic-equivalent-medium vertical phase increment of the P-wave transmissivity and γ_p is its dynamic-equivalent-medium attenuation coefficient. Because these results are valid for any relationship between the wavelength and the correlation length of the medium fluctuations, we call both quantities 'dynamic-equivalent-medium parameters'.

In order to obtain the (dynamic-equivalent-medium) vertical-phase increment ψ_s and attenuation coefficient γ_s of the SV-wave transmissivity, we must repeat the derivation for the P-wave transmissivity, but now, however, with a new choice of boundary conditions:

$$d_p(0) = 0, \quad d_s(0) = 1, \quad u_p(L) = 0, \quad u_s(L) = 0. \tag{5.53}$$

In a similar manner we arrive at equations (5.37) for SV-waves.

6 Frequency-Dependent Properties of Stratigraphic Filtering

In the previous chapters a set of generalized O'Doherty-Anstey (ODA) formulas has been established. These formulas, describing approximately the effect of stratigraphic filtering, are used here to study most important features of this effect in the frequency range. These features are (i) *the frequency-dependent velocity anisotropy*, (ii) *the frequency-dependent shear-wave splitting*, and (iii) *the dispersion and attenuation of the time-harmonic transmissivity*. They characterize the frequency-domain behavior of generalized primaries of seismic events.

This chapter concentrates on the transmissivity in the case of stationary medium fluctuations. The corresponding set of the generalized ODA formulas is given by equations (4.18)- (4.22) and (5.36)-(5.37).

An important advantage of the generalized ODA formulas is their validity for the whole-frequency range, i.e., the validity of the results is independent of the relationship between the wavelength and the correlation length of the medium fluctuations. The attenuation coefficients and vertical-phase increments given by these formulas (and all other attributes derived from them) can be looked upon as describing transmitted plane quasi P-, quasi SV- and SH waves that are propagating in a frequency-dependent dynamic-equivalent medium. It will be shown that the low-frequency limits of the generalized ODA formulas for the phase increments ψ_P, ψ_{SV} and ψ_{SH} yield results equal to those obtained by Backus averaging (Backus, 1962). Also the high-frequency limits of these formulas are equivalent to the solutions of geometrical optics.

The phase velocity $c_{P,SV,SH}$ (in a dynamic-equivalent-medium) can be expressed as:

$$c_{P,SV,SH}(\omega,p) = 1/\sqrt{p^2 + \psi^2_{P,SV,SH}/\omega^2}. \tag{6.1}$$

This is a fundamental attribute in the following consideration concerning the frequency dependence of the velocity anisotropy and shear-wave splitting. The p-

dependence of the phase velocity for different frequencies describes the frequency-dependent velocity anisotropy.

For the analysis of the attenuation of the time-harmonic transmissivities, which decay like $\exp(-\gamma L)$ due to multiple scattering within the medium, along with the attenuation coefficient γ we will also use the *reciprocal quality factor* Q^{-1}. Following the notations of Johnston and Toksöz (1981) we define this factor formally in the following way:

$$Q^{-1} = 2\gamma/k. \tag{6.2}$$

In this expression k is the wavenumber of the corresponding plane wave in the homogeneous reference medium.

6.1   Low-Frequency Asymptotic Solution

The low-frequency limits of expressions (5.36), (5.37) and (4.19) yield the following approximations for the vertical-phase increments and attenuation coefficients:

$$
\begin{aligned}
\psi_P^{low} &= \kappa_a + \omega A_P, \\
\gamma_P^{low} &= \omega^2 \int_0^\infty d\xi \left[B_P(\xi) + B_{BB}(\xi) + B_{DD}(\xi) \right], \\
\psi_{SV}^{low} &= \kappa_b + \omega A_{SV}, \\
\gamma_{SV}^{low} &= \omega^2 \int_0^\infty d\xi \left[B_{SV}(\xi) + B_{BB}(\xi) + B_{DD}(\xi) \right], \\
\psi_{SH}^{low} &= \kappa_b + \omega A_{SH}, \\
\gamma_{SH}^{low} &= \omega^2 \int_0^\infty d\xi B_{SH}(\xi).
\end{aligned}
\tag{6.3}
$$

Substituting the expressions for $\psi_{P,SV,SH}^{low}$ into equation (6.1), we obtain the low-frequency approximations for the absolute values of the phase velocities of the qP-, qSV- and SH- waves:

$$
\begin{aligned}
c_P^{low} &= a(1 - aA_P X), \\
c_{SV}^{low} &= b(1 - bA_{SV} Y), \\
c_{SH}^{low} &= b(1 - bA_{SH} Y).
\end{aligned}
\tag{6.4}
$$

The explicit formulas for the quantities A_P, A_{SV} and A_{SH} are given in equations (4.20) and (5.40). Equations (6.4) describe the wavefield kinematics in the *effective transversely-isotropic elastic medium*, where the elastic moduli are defined by Backus averaging.

This can be proven by the following steps: (i) express the Backus-averaged elastic moduli in terms of the medium fluctuations (5.3) and neglect all terms higher than second order in the fluctuations, (ii) substitute these moduli into the system of equations describing the components of the wave vectors of P-SV and SH- waves in a transversely isotropic medium (see, e.g., equations (6) - (8) in Berryman, 1979); (iii) solve this system of equations for the vertical component of the wave vector; (iv) substitute the resulting three solutions for the vertical components of the wavevectors of qP-, qSV- and SH- waves instead of $\psi_{P,SV,SH}$ in formula (6.1); (v) perform a Taylor expansion of the square roots and neglect again all terms higher than the second order in the fluctuations. These calculations will yield the absolute values of the phase velocities exactly coinciding with c_P^{low}, c_{SV}^{low} and c_{SH}^{low} respectively.

Therefore, in the low-frequency limit the generalized ODA formulas describe the usual *effective-medium transverse isotropy*.

The quantities γ_P^{low}, γ_{SV}^{low} and γ_{SH}^{low} describe the low-frequency asymptotic behavior of the attenuation coefficients, which are proportional to the squared frequency. This is typical for *Rayleigh scattering* in 1-D media.

6.2   High-Frequency Asymptotic Solution

The high-frequency limits of expressions (4.19), (5.36) and (5.37) can be obtained by using well-known properties of the Fourier transform. These high-frequency expressions strongly depend on the continuity of the auto- and crosscorrelation functions (here summarily referred to as $\Phi(\xi)$) of the medium fluctuations. If $\Phi(\xi)$ is continuous but its first derivative is not then the cosine transform will decay in the high-frequency limit like $1/\Omega^2$, where Ω is the argument of the Fourier transform (i.e., $\Omega = 2\kappa_a, 2\kappa_b, \kappa_-, \kappa_+$). Such a behavior is typical for correlation functions of a discretely layered medium or for an exponential correlation function. This in turn implies that for such a medium the attenuation coefficient is approximately constant in the high-frequency range.

However, the medium can also have other correlation functions, which are n-times differentiable for all lags ξ. This leads to a $1/\omega^{2+n}$–decay of the cosine transform and, therefore, to a decrease of the attenuation coefficient like $1/\omega^n$.

Therefore, in both cases one observes a non-increasing behavior of the attenuation coefficient in the limit of *infinitely high frequencies*.

The high-frequency limit of the sine transform of the correlation function given by $\Phi(\xi)$ can be calculated explicitly. It is $\Phi(0)/\Omega$. This leads to a simple expression for the absolute values of the phase velocities:

$$
\begin{aligned}
c_P^{high} &= a\left(1 - \sigma_{aa}^2 \frac{1 - 3p^2 a^2/2}{1 - p^2 a^2}\right), \\
c_{SV,SH}^{high} &= b\left(1 - \sigma_{bb}^2 \frac{1 - 3p^2 b^2/2}{1 - p^2 b^2}\right).
\end{aligned}
\tag{6.5}
$$

These relations are in full agreement with the result of ray theory:

$$
c^{ray} = \sqrt{L^2 + x^2}/\tau,
$$

where L and x are the total vertical and horizontal distances defining a ray and τ denotes the total travel time. In order to obtain the above formula we approximate the inhomogeneous medium by a stack of many homogeneous layers of equal thickness h. Then the quantities x and τ are

$$
\begin{aligned}
x &= h\sum_i \tan \vartheta_i, \\
\tau &= h\sum_i 1/(\alpha_i \cos \vartheta_i),
\end{aligned}
$$

where ϑ_i and α_i are the angles of incidence and (compressional/shear wave) velocities in each layer, respectively. The sums can be replaced by spatial averages (which are equal, by definition, to ensemble averages for ergodic random media) multiplied by the number of layers. Using this and remembering that the horizontal slowness p is constant, we obtain

$$
\begin{aligned}
x &= Lp\left\langle \frac{\alpha}{\sqrt{1 - p^2 \alpha^2}} \right\rangle, \\
\tau &= L\left\langle \frac{1}{\alpha\sqrt{1 - p^2 \alpha^2}} \right\rangle,
\end{aligned}
$$

where $\alpha = \alpha(z)$ is the depth-dependent compressional or shear velocity. Substituting these relations into the formula for c^{ray}, performing all necessary algebra and neglecting higher than second-order powers in the fluctuations, we obtain results which exactly coincide with formulas (6.5).

6.3 Whole-Frequency Domain Solution

To illustrate the main features of stratigraphic filtering in the whole-frequency domain we specify now the statistical properties of the random medium more precisely. We consider a medium with an exponential correlation functions $\Phi(\xi) = \sigma^2 \exp(-\xi/l)$, where l is the correlation length. Such a medium is of much practical interest. We assume that functions $\Phi(\xi)$, with different values for the variance σ^2, approximate well the autocorrelation of the velocities- and the density fluctuation as well as their crosscorrelations.

With this choice of the medium statistics, in formulas (4.19), (5.36) and (5.37) we must replace the Fourier sine- and cosine transforms of the corresponding correlation functions $\Phi(\xi)$ by the following analytic expressions, respectively:

$$\sigma^2 \frac{l^2 \Omega}{1 + l^2 \Omega^2} \quad \text{and} \quad \sigma^2 \frac{l}{1 + l^2 \Omega^2}. \qquad (6.6)$$

After this substitution we obtain the following simple formulas for the (dynamic-equivalent medium) attenuation coefficients and vertical-phase increments of the

P-, SV- and SH-wave transmissivities :

$$\psi_P = \kappa_a + \omega A_P -$$
$$- \omega^2 l^2 \left[B_P(0)\frac{2\kappa_a}{1+4l^2\kappa_a^2} + B_{BB}(0)\frac{\kappa_-}{1+l^2\kappa_-^2} + B_{DD}(0)\frac{\kappa_+}{1+l^2\kappa_+^2} \right],$$
$$\gamma_P = \omega^2 l \left[B_P(0)\frac{1}{1+4l^2\kappa_a^2} + B_{BB}(0)\frac{1}{1+l^2\kappa_-^2} + B_{DD}(0)\frac{1}{1+l^2\kappa_+^2} \right],$$
$$\psi_{SV} = \kappa_b + \omega A_{SV} -$$
$$- \omega^2 l^2 \left[B_{SV}(0)\frac{2\kappa_b}{1+4l^2\kappa_b^2} - B_{BB}(0)\frac{\kappa_-}{1+l^2\kappa_-^2} + B_{DD}(0)\frac{\kappa_+}{1+l^2\kappa_+^2} \right], \quad (6.7)$$
$$\gamma_{SV} = \omega^2 l \left[B_{SV}(0)\frac{1}{1+4l^2\kappa_b^2} + B_{BB}(0)\frac{1}{1+l^2\kappa_-^2} + B_{DD}(0)\frac{1}{1+l^2\kappa_+^2} \right],$$
$$\psi_{SH} = \kappa_b + \omega A_{SH} - \omega^2 l^2 B_{SH}(0)\frac{2\kappa_b}{1+4l^2\kappa_b^2},$$
$$\gamma_{SH} = \omega^2 l B_{SH}(0)\frac{1}{1+4l^2\kappa_b^2}.$$

In the following we will use these formulas and provide several examples showing the frequency and angle dependencies of the phase velocities (given by the substitution of equations (6.7) into equation (6.1)), of the shear-wave splitting (given by substituting equations (6.7) and (6.1) into equation (6.8)) and of the quality factors of the transmissivity (given by substituting equations (6.7) into equation (6.2)).

We also compare the theoretical curves with results of numerical simulations of exact wavefields. For these we have chosen the values $a = 4000m/s$, $b = 2850m/s$, $\varrho = 2.5g/cm^3$, $l = 5m$, $\sigma_{aa} = 0.15$, $\sigma_{bb} = 0.14$, $\sigma_{\varrho\varrho} = 0.12$, $\sigma_{ab}^2 = 0.021$, $\sigma_{a\varrho}^2 = 0.018$ and $\sigma_{b\varrho}^2 = 0.017$. A particular realization of the sonic log in such a random medium is shown in Figure 2. All frequency-dependent transmissivity attributes are plotted in the frequency range $0 - 500Hz$. The corresponding ranges of the non-dimensional frequencies $\omega l/a$ and $\omega l/b$ are 0 to about 4 and 0 to about 5.5, respectively.

The resulting phase velocity of the P-wave transmissivity is computed (Figure 12) for four different incidence angles $\vartheta = 0, 30, 45$ and 60 degrees (we define the angle of incidence for the homogeneous reference medium as $\sin\vartheta = pa$ for the incident P-wave and as $\sin\vartheta = pb$ for the incident S-wave). All four curves start on the vertical axes (where the frequency is equal to zero) at velocity values slightly different from those given by Backus averaging (because we neglect the effect of terms containing third and higher powers of fluctuations). With increasing frequency they tend to the values resulting from geometrical optics. For comparison the straight line denotes the RMS velocity (for more details see

Hubral and Krey, 1980). The anisotropy is revealed by the vertical separation of
the curves for a fixed frequency. A strong frequency dependence of the anisotropy
can be easily observed. It can be noted that in the frequency range of seismic
exploration (e.g., from 20 to 150 Hz) the anisotropy differs significantly from
that described by Backus averaging.

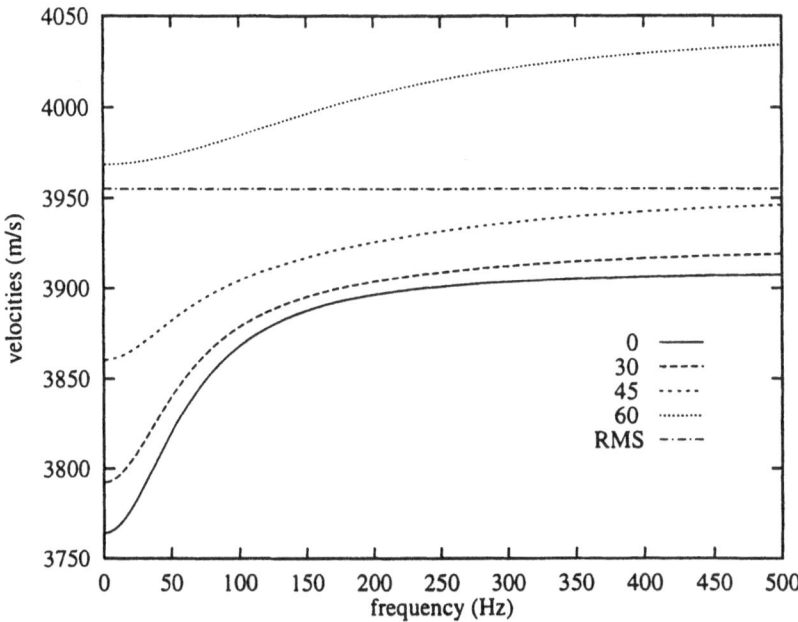

Fig. 12. Frequency and angle dependence of the phase velocity for the P-wave transmissivity. The parameters of the medium are $a = 4000m/s$, $b = 2850m/s$, $\varrho = 2.5g/cm^3$, $l = 5m$, $\sigma_{aa} = 0.15$, $\sigma_{bb} = 0.14$, $\sigma_{\varrho\varrho} = 0.12$, $\sigma_{ab}^2 = 0.021$, $\sigma_{a\varrho}^2 = 0.018$ and $\sigma_{b\varrho}^2 = 0.017$. (from Shapiro and Hubral, 1996).

Figure 13 shows the dynamic-equivalent-medium reciprocal quality factor Q^{-1}
characterizing the attenuation of the time-harmonic P-wave transmissivity. For
the given example we can observe that in the low-frequency range the following
holds: the smaller the incidence angle the larger is Q^{-1}. However, in the high-
frequency range this dependence is inverted.

The next two figures show phase velocities (Figure 14) and reciprocal quality
factors (Figure 15) for the transmissivity of the SV-wave. As we study only
under-critical incidence and the parameters σ^2/X^2 and σ^2/Y^2 are assumed to
be small, we have plotted these quantities only for angles up to 30°. Again we can
see quite complicated frequency and angle dependencies of the studied quantities.
The phase velocity reveals a significant frequency-dependent anisotropy. It is

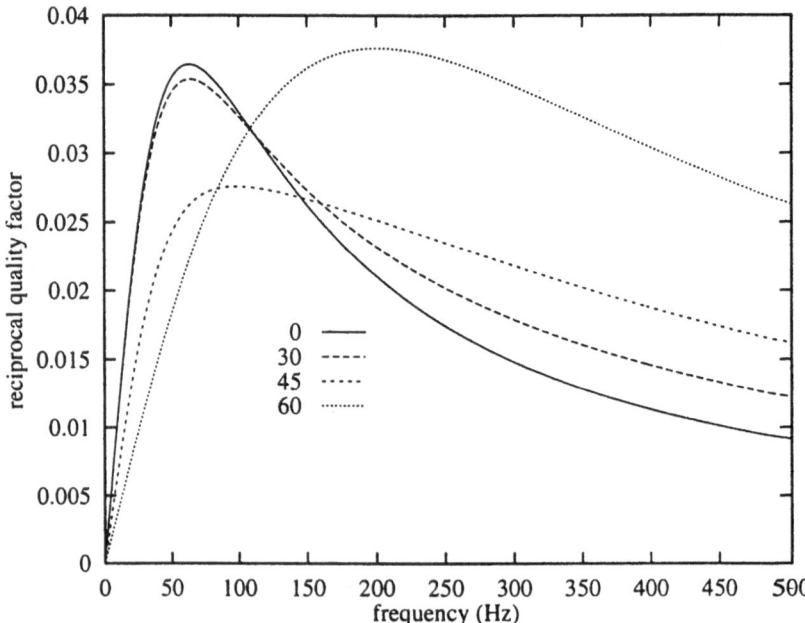

Fig. 13. Frequency and angle dependence of the (dynamic-equivalent medium) quality factor Q^{-1} of the P-wave transmissivity. The medium parameters are the same as in Figure 12. (from Shapiro and Hubral, 1996).

interesting to observe that in the intermediate frequency range (50 - 200 Hz) the phase velocity reaches its maximum value for oblique incidence. Due to the SV-P-SV mode conversion, this value is even larger than the geometrical-optics limit.

The next two illustrations (Figures 16 and 17) show phase velocities and recipro- cal Q-factors of the SH-wave. Again, a significant frequency-dependent anisotropy of the velocity and of the attenuation can be observed. For example, at zero fre- quency the difference between velocities for the angles of incidence of 0 and 45 degrees is of the order of 4 per cent. At the frequency of 100 Hz it is of the order of 1.5 per cent only.

For this example of the stratification we provide below some results of numer- ical simulations of exact wavefields. We applied the classical *matrix-propagator (Thomson-Haskell) method*. For the numerical simulation an exponential medium was used with a thickness of each elementary homogeneous layer equal to 1m. We assumed for simplicity that the velocities and the density are completely correlated. Therefore, the fluctuations of the shear and density logs are just pro- portional to those of the sonic log. They are, however, characterized by different variances. The formulas (5.36), (5.37) and (6.7), however, allow for any degree

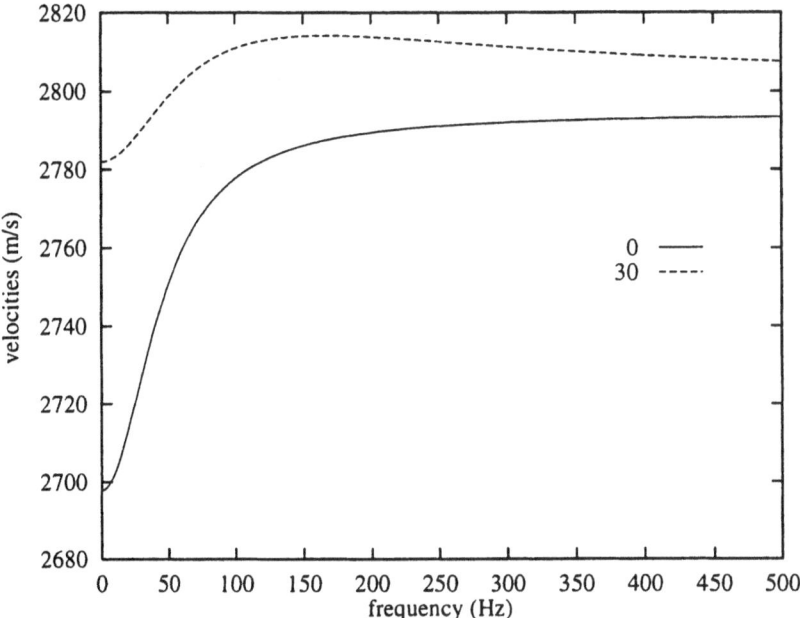

Fig. 14. Frequency and angle dependence of the phase velocity for the SV-wave transmissivity. The medium parameters are the same as for Figure 12.

of correlation between the logs.

We would like to remark additionally that this particular example of the inhomogeneous medium falls far short of being the most convenient one for a demonstration of the validity of the described theory. The variances of the velocities and density are very high (usually for a real sonic log they are smaller than ten per cent). This restricts the validity domain (see condition (5.46)) of L to the narrow interval $\lambda < L < 200$m, where λ is the wavelength. Below we compare the numerically computed exact time-harmonic transmissivity and the analytical one for $L = 200$m. This we do for arbitrary frequencies, including the case $L < \lambda$, and violating, therefore, the lower boundary of the interval (5.46). This comparison for the transient transmissivity, as shown in the next chapter, will be made for $L = 1000m$ and $L = 2000m$ violating, therefore, the upper boundary of the interval (5.46). It will be observed that even though the values of L are outside the range (5.46), the theoretical results are still in good agreement with the numerical ones.

Due to the random character of the medium fluctuations all numerical results shown here are oscillating. The theoretical curves have been computed with a smooth exponential correlation function, which characterizes the statistical properties of a hypothetical ensemble of realizations of the random medium. This

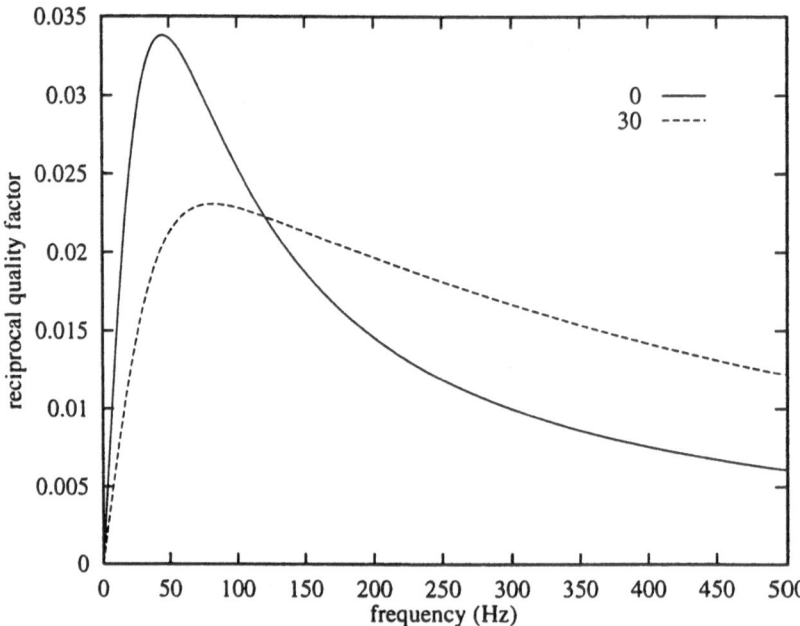

Fig. 15. Frequency and angle dependence of the dynamic-equivalent-medium quality factor Q^{-1} of the SV-wave transmissivity. The medium parameters are the same as for Figure 12.

correlation function does not have the small-scale fluctuations always revealed by the correlation functions of the single realizations. Therefore, the theoretical curves cannot predict all oscillations in the numerical results. Thus, when speaking here about the agreement between numerical and theoretical curves, we imply that the theoretical curves represent only a smooth version of the numerically-computed curves.

The most complicated part of the numerical computation is the simulation of the phase velocity. For this an unwrapped (i.e. continuous) phase function of the time-harmonic transmissivity is needed. However, procedures for the phase reconstruction of the numerically computed transmissivity in a random medium are principally unstable. Fortunately, in the case of a not very thick layered stack (e.g., $L = 200m$) more or less satisfactory results can still be obtained.

In Figure 18 the frequency dependence of the phase velocity of the P-wave transmissivity is shown for the angle of incidence of 30°. The isolated points (diamonds) denote numerical results for the above model with $L = 200m$. The dashed line on the other hand presents the analytical result (obtained by the substitution of ψ_p from equation (6.7) into equation (6.1)). A good agreement can be observed. The next plot (Figure 19) shows such a comparison for the

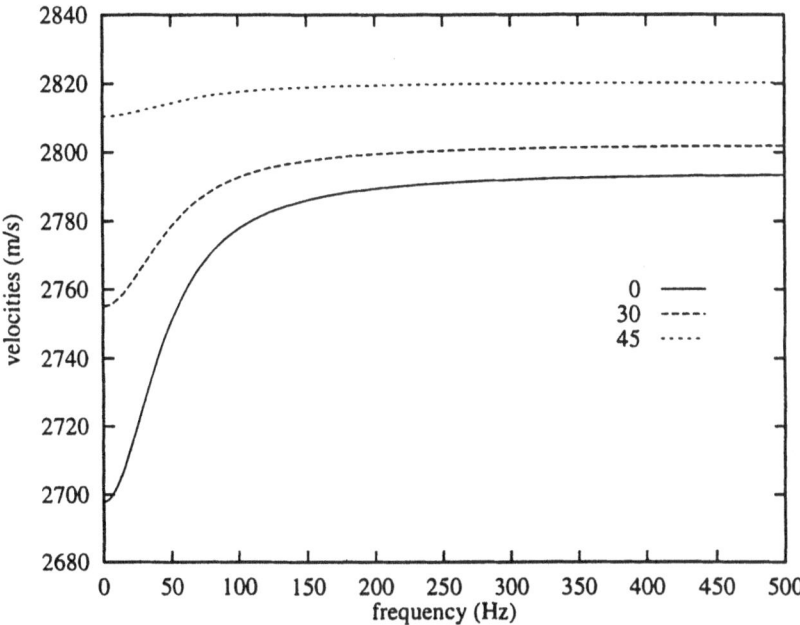

Fig. 16. Frequency and angle dependence of the phase velocity for the SH-wave trans-missivity. The medium parameters are the same as for Figure 12.

angle of incidence equal to 45°. The extremely high values of the velocity close to $315Hz$ and $360Hz$ are due to an incomplete phase unwrapping. They are thus nothing but a purely numerical effect. Finally, Figure 20 shows the numerically and analytically calculated phase velocity of the SV-wave transmissivity for the angle of incidence equal to 30°. Again close to $230Hz$ and $300Hz$ we observe a purely numerical effect due to incomplete phase unwrapping. However, in general, a good overall agreement between analytical and numerical results is evident.

For the attenuation coefficients the results of exact numerical computations (i.e. $ln|T_{p,s}(\omega)|/L$) are likewise compared with those of the analytical solutions γ_p and γ_s from (6.7). The comparison is shown in Figures 21 and 22. They show the attenuation coefficients for the P-wave transmissivity (angle of incidence is 45°) and the SV-wave transmissivity (angle of incidence is 30°). At high frequencies the attenuation coefficient reaches a constant level. Again an agreement between theoretical and numerical results can be observed. Because of very strong medium fluctuations (the thickness L is just equal to the upper boundary of the validity range (5.46)) the theoretical results tend to overestimate the attenuation coefficient for frequencies higher than 400 Hz. This effect will disappear if L or ε decreases.

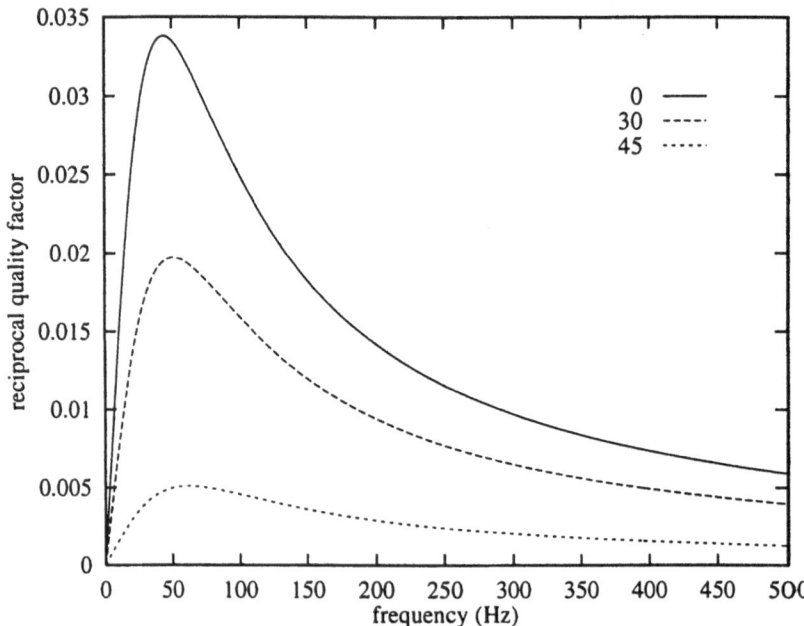

Fig. 17. Frequency and angle dependence of the dynamic-equivalent-medium quality factor Q^{-1} of the SH-wave transmissivity. The medium parameters are the same as for Figure 12.

The above examples show that there exists a need in seismic stratigraphic investigations to use in such complicated (but realistic) media, as investigated here, the frequency dependence of the considered wavefield attributes. They can be explained in terms of the simpler *dynamic-equivalent medium wavefield attributes*. For the case of exponential correlation functions of the medium fluctuations the *dynamic-equivalent medium* (that was defined above with respect to characterizing the P-, SV- and SH-waves transmissivities) is in our approximation entirely described by *ten* real constants. These are the correlation length l, averaged velocities and density a, b and ϱ and six variances and covariances $\sigma_{aa}^2, \sigma_{ab}^2, \sigma_{bb}^2, \sigma_{a\varrho}^2, \sigma_{b\varrho}^2$ and $\sigma_{\varrho\varrho}^2$. It is interesting to note for comparison that in the general case of a *viscoelastic transversely-isotropic medium* eleven parameters are needed. These are the real and imaginary parts of the five viscoelastic frequency-dependent moduli and the density.

6.4 Frequency-Dependent Shear-Wave Splitting

Formulas (4.19, 4.20, 5.36, 5.37) and (5.40) show that the anisotropy due to multilayering can reach the order of $10^3\varepsilon^2$ per cent, where ε is the order of

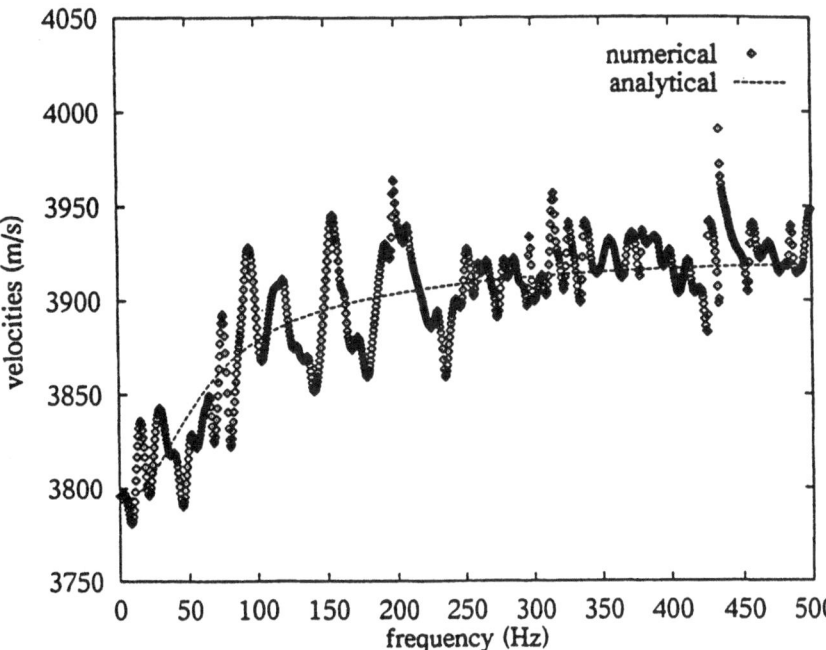

Fig. 18. Comparison of numerical and theoretical estimations of the frequency dependence of the P-wave-transmissivity phase velocity. Angle of incidence is 30 degrees. The medium parameters are the same as for Figure 12. For the numerical computation $L = 200m$. (from Shapiro and Hubral, 1996).

the medium fluctuations. However, for not very large angles of incidence the effects of anisotropy are smaller. This is also the case for the shear-wave splitting observed for angles of incidence smaller then 45 degrees. Usually, it is a quantity of the order of $10^2\varepsilon^2$ per cent or smaller. Moreover the shear-wave splitting has a tendency to decrease for increasing frequencies. Indeed, with knowledge of the phase velocities of SV- and SH-waves the shear-wave splitting can be studied. This frequency-dependent shear-wave splitting can be characterized by the following measure:

$$S(\omega, p) = \frac{c_{SV}(\omega, p) - c_{SH}(\omega, p)}{c_{SV}(\omega, p)} \approx \frac{c_{SV}(\omega, p) - c_{SH}(\omega, p)}{b}. \tag{6.8}$$

In fact the quantity S describes the frequency dependence of the so-called shear-wave birefringence (Winterstein, 1990). In this book, however, no difference is

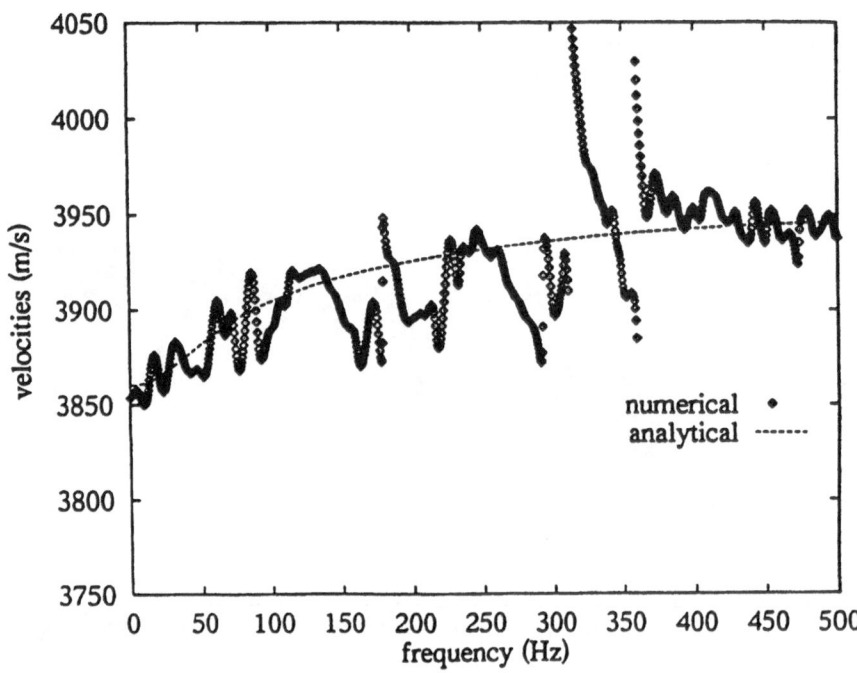

Fig. 19. The same as Figure 18, but the angle of incidence is 45 degrees. (from Shapiro and Hubral, 1996).

made between the two terms *birefringence* and *splitting*.

Substituting equations (6.4) into equation (6.8) we obtain for the shear-wave splitting in the low-frequency limit:

$$S(\omega, p)^{low} \approx bY(A_{SH} - A_{SV}). \tag{6.9}$$

This is the shear-wave splitting in the effective transversely isotropic medium with elastic moduli given by the Backus averaging. In the high-frequency limit the substitution of equations (6.5) into equation (6.8) immediately yields the physically obvious result $S(\omega, p)^{high} = 0$.

Thus, we have already arrived at the conclusion that the shear-wave splitting must be frequency dependent. In order to analyze it in the total frequency range we assume the medium to be exponential and substitute the expressions for the

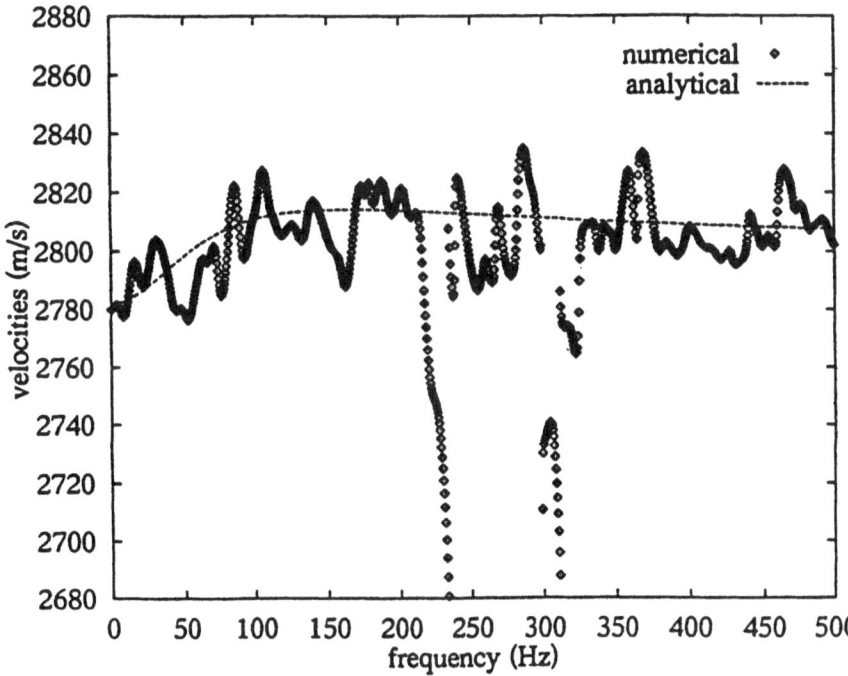

Fig. 20. The same as Figure 18, but the SV-wave transmissivity is analyzed. (from Shapiro and Hubral, 1996).

quantities ψ_{SV} and ψ_{SH} into equation (6.8). We obtain:

$$S(\omega,p) \approx bY\{A_{SH} - A_{SV} + \frac{2\omega l^2 \kappa_b}{1 + 4l^2 \kappa_b^2}(B_{SV}(0) - B_{SH}(0)) \\ - \omega l^2[\frac{\kappa_+}{1 + l^2 \kappa_+^2}B_{DD}(0) - \frac{\kappa_-}{1 + l^2 \kappa_-^2}B_{BB}(0)]\}. \tag{6.10}$$

Taking into account that the quantities $\kappa_a, \kappa_b, \kappa_+$ and κ_- are proportional to the frequency, we immediately observe that expression (6.10) demonstrates a frequency dependence of the shear-wave splitting.

Figure 23 shows plots of the shear-wave splitting versus frequency for two different angles of incidence. It can be seen clearly that the values of the shear-wave splitting in the seismic frequency range differ from those at zero frequency. This frequency dependence of the shear-wave splitting is very significant. For instance, in the model considered in the case of the angle of incidence equal to 30° the shear-wave splitting in the static limit is approximately 1 per cent. However, at 40 Hz it is approximately 0.7 per cent. It is clear that such a frequency de-

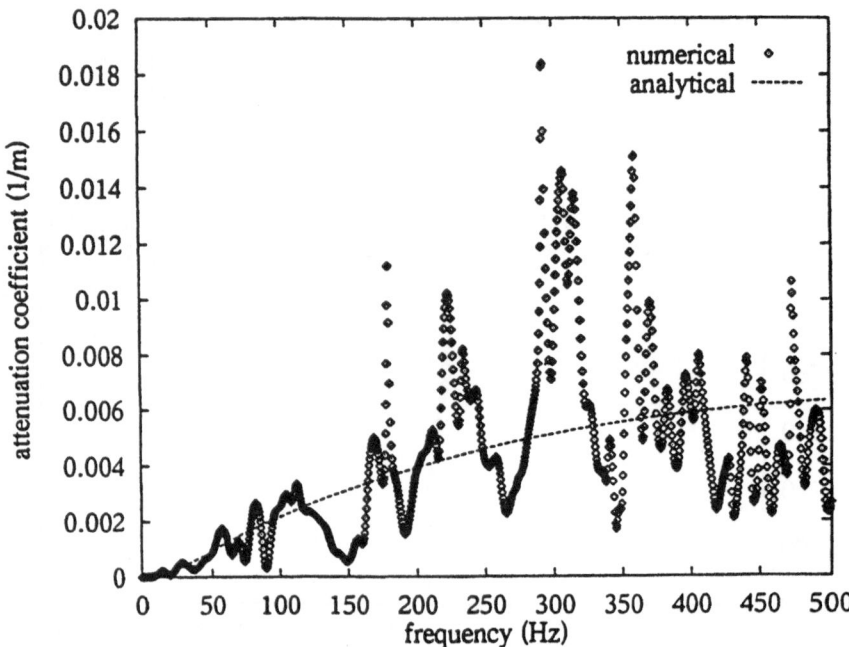

Fig. 21. Comparison of numerical and theoretical estimations of the frequency-dependent P-wave-transmissivity attenuation coefficient. Angle of incidence is 45 degrees. The medium parameters are the same as in Figure 12. (from Shapiro and Hubral, 1996).

pendence can have consequences for the interpretation of shear-wave splitting. A general decrease of the shear-wave splitting with frequency is observed. It must be taken into account when analyzing seismic data collected in different frequency ranges: surface seismic, VSP, cross-hole observation and laboratory measurements.

Werner and Shapiro (1997) considered the frequency-dependent shear-wave splitting in *multilayered media with intrinsic anisotropy in the form of the transverse isotropy with the vertical symmetry axes.* They showed that in such a situation the shear-wave splitting depends on the frequency only if medium parameters fluctuate. In the absence of fluctuations it depends only on the incidence angle. The contribution of the intrinsic anisotropy is a shift of the phase velocities and shear-wave splitting by frequency-independent amounts that are controlled by the magnitude of the intrinsic anisotropy. Therefore, in contrast to the case where the anisotropy is caused by isotropic multilayering only (as in the situations considered above), the shear-wave splitting does not vanish at high frequencies. The contributions of both anisotropy effects on the shear-wave splitting are in the **first approximation additive.**

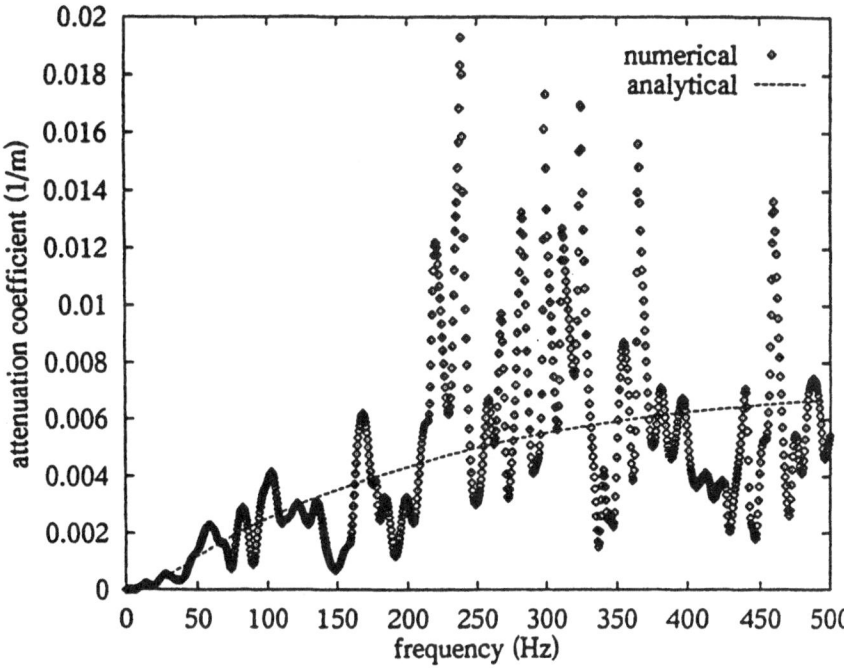

Fig. 22. The same as Figure 21 but the SV-wave transmissivity is analyzed. Angle of incidence is 30 degrees. (from Shapiro and Hubral, 1996).

6.5   Fluctuations of the Attenuation Coefficient

Stratigraphic filtering turns the self-averaged quantities, the attenuation coefficient and the vertical-phase increment of the transmissivity, into non-random values. This is, however, true for an infinite stack of layers only. In practice a layered medium is of course always finite. This means that both quantities do indeed fluctuate when computed for finite individual realizations of a random medium.

For simplicity let us consider the fluctuations of the attenuation coefficient γ_{ac} in a randomly layered fluid. Up to the second order in the small parameter ε the variance of the attenuation coefficient σ_γ^2 can be directly obtained from the averaged square of the logarithm of the amplitude-like term r_s given in equation (4.12). Again we estimate the fluctuations in the limit of a large medium. This

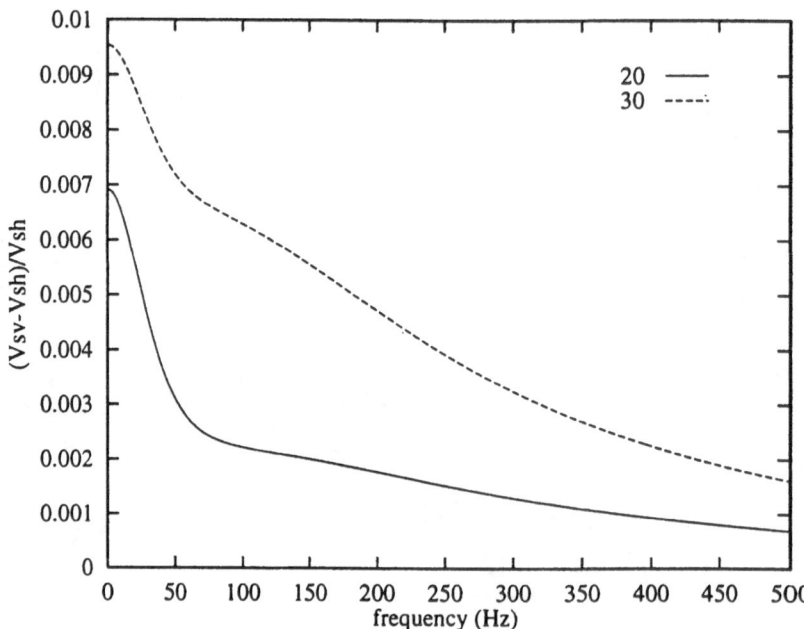

Fig. 23. Frequency and angle dependence of the shear-wave splitting S. The medium parameters are the same as for Figure 12.

means that we perform an asymptotic expansion of

$$\sigma_\gamma^2 = \left\langle \left[\frac{\ln r_s(L)}{L} \right]^2 \right\rangle \tag{6.11}$$

for large L. Substitution of equation (4.15) and performing the same derivation as outlined after equations (4.16) yields

$$\sigma_\gamma^2 = \frac{\omega^2}{L} \int_0^\infty d\xi \, B_{ac}(\xi) \cos(2\kappa_{ac}\xi) + o(1/L). \tag{6.12}$$

By comparing this relation with the attenuation coefficient (4.18), we can write its relative standard deviation in the following way:

$$\frac{\sigma_\gamma}{\gamma} = \sqrt{\frac{1}{\gamma L}}. \tag{6.13}$$

This quantity vanishes for large media like $1/\sqrt{L}$ because of the self-averaged property of the attenuation coefficient.

By an analogous derivation, it can be shown that the standard deviation of the phase increment ψ also vanishes for large media like $1/\sqrt{L}$. This implies that the standard deviation of the total phase will increase like $\sqrt{L}$. It may appear that this fact makes the present approximation useless, because the phase is an angle, and therefore, should be bounded by 2π. However, this is not true. First, our error estimation refers to the *unwrapped phase*, which is an increasing function of the traveldistance and is equal to $\omega \times$ *traveltime*. Therefore, the relative standard deviation of this quantity is decreasing like $1/\sqrt{L}$. Secondly, not the phase, but only the phase increment has a sound physical sense, because it controls the phase and group velocities. The standard deviations of the velocities vanish like $1/\sqrt{L}$ in our approximation.

6.6   Inversion for Statistics of Stratifications

The properties of the stratigraphic filtering can be used for an inversion of the seismic transmission and reflection events to estimate the statistical character-istics of targets (described by a stack of layers) or the overburden.

In the paper of Shapiro and Zien (1993) a *statistical inversion* procedure was presented for the case of a P-wave vertically incident on a multilayered medium with constant density. It was indicated how the (measured) frequency-dependent attenuation coefficient could be inverted for the correlation length of the random-medium velocity fluctuations. The inversion in the cited paper was successfully performed on an example of a synthetic data set obtained by White et al., (1990) for real sonic logs. Using the same inversion procedure extended, however, to the case of vertically incident P-waves in a medium with a variable compres-sional velocity and density, Zien (1993) processed real data. He extracted the frequency-dependent attenuation coefficient from a zero-offset Vertical Seismic Profiling (VSP) data set acquired in the North Sea. Assuming that the sedimen-tary overburden in the depth range 640-2698m is characterized by an exponential correlation function of the density and compressional velocity fluctuations, Zien estimated their correlation lengths from the attenuation coefficient and compared the results with values of the correlation lengths obtained directly from sonic- and density logs. This comparison is shown in Figure 24. It can be observed that the results of seismic inversion agree well with the log measurements.

Formulas (5.36) and (5.37) allow not only for a generalization of such an in-version to the case of oblique incidence in a solid. They also provide additional

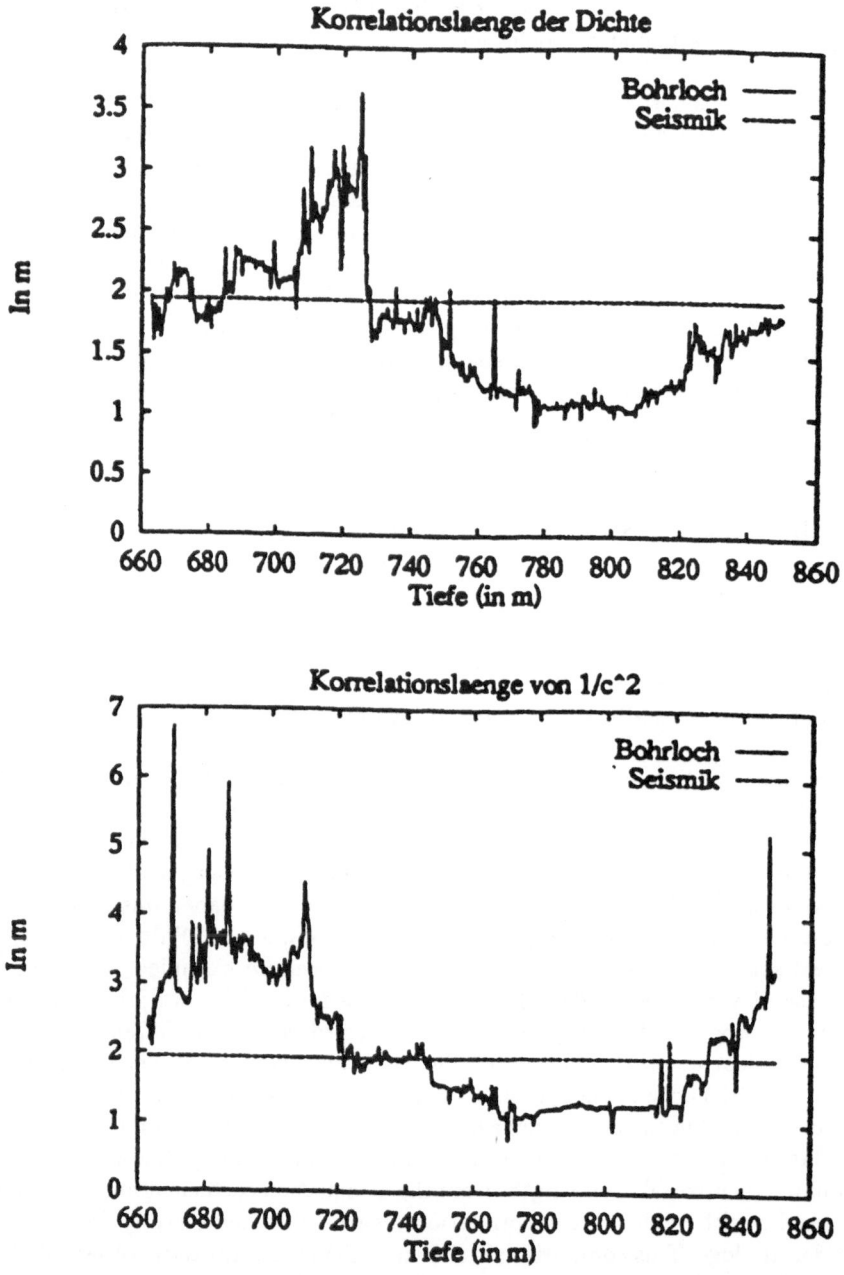

Fig. 24. Correlation length (German: Korrelationslaenge) of the density (top) and the squared slowness (bottom) fluctuations estimated from log data (solid lines) in comparison with an estimation of the correlation distance obtained by the inversion of seismic data (dashed straight lines). The horizontal axis shows the depth (German: Tiefe) in m. The vertical axis shows the correlation length (in m.) This figure has been taken, with kind permission, from Zien (1993, p.77)

possibilities for inversion taking the angle dependence of the attenuation coefficients into account. For example, it would be possible by inversion to find the statistics of the density and the compressional and shear velocities separately. Another potential for inversion can be found in using directly the pulse signature or its phase behavior.

7 Transient Transmissivity

As we have seen in Section 3.4 the transient transmissivity (i.e. the transmitted signal caused by the incident δ-pulse plane wave) is composed of the generalized primary and the coda. In this chapter we consider the properties of the generalized primary in the case of obliquely incident P-, SV- and SH-waves.

7.1 Coherent and Incoherent Parts

Expressions (4.19), (5.36), and (5.37) for the phase increment $\psi_{P,SV,SH}$ and attenuation coefficient $\gamma_{P,SV,SH}$ of the time-harmonic transmissivities (5.35) and (4.22) allow us to analyze the transient transmissivity. The transient transmissivity $\tilde{T}_{P,SV,SH}(t)$ for an incident delta–pulse P-, SV- or SH plane wave can be calculated by the integration (which is an inverse Fourier transform):

$$\tilde{T}_{P,SV,SH}(t) = \frac{1}{2\pi} \int_{-\infty}^{\infty} d\omega \, T_{P,SV,SH}(\omega). \qquad (7.1)$$

Generally, the transient transmissivity can be expressed in the following way:

$$\begin{aligned}
\tilde{T}_{P,SV;SH}(t) &= \\
&= \delta(t - px - \tilde{\psi}_{P,SV,SH}^{low} L) * S_{1P,SV,SH}(t) * S_{2P,SV,SH}(t) * S_{3P,SV,SH}(t),
\end{aligned} \qquad (7.2)$$

where the asterisk denotes convolution in time. The quantities $\tilde{\psi}_{P,SV,SH}^{low}$ are defined by

$$\psi_P^{low} = X/a + A_P, \quad \psi_{SV,SH}^{low} = Y/b + A_{SV,SH}. \qquad (7.3)$$

The functions $S_{2SH}(t)$ and $S_{3SH}(t)$ are equal to the delta-function $\delta(t)$ and the functions $S_{1P,SV,SH}(t), S_{2P,SV}(t)$ and $S_{3P,SV}(t)$ are

$$S_{1P}(t) = \frac{1}{2\pi} \int_{-\infty}^{\infty} d\omega \, \exp\left[-L\omega^2 \hat{B}_P(2\kappa_a) - i\omega t\right],$$

$$S_{1SV,SH}(t) = \frac{1}{2\pi} \int_{-\infty}^{\infty} d\omega \, \exp\left[-L\omega^2 \hat{B}_{SV,SH}(2\kappa_b) - i\omega t\right],$$

$$S_{2P}(t) = \frac{1}{2\pi} \int_{-\infty}^{\infty} d\omega \, \exp\left[-L\omega^2 \hat{B}_{BB}(\kappa_-) - i\omega t\right], \qquad (7.4)$$

$$S_{2SV}(t) = \frac{1}{2\pi} \int_{-\infty}^{\infty} d\omega \, \exp\left[-L\omega^2 \hat{B}_{BB}(-\kappa_-) - i\omega t\right],$$

$$S_{3SV}(t) = S_{3P}(t) = \frac{1}{2\pi} \int_{-\infty}^{\infty} d\omega \, \exp\left[-L\omega^2 \hat{B}_{DD}(\kappa_+) - i\omega t\right].$$

The functions $\hat{B}_{P,SV,SH,BB,DD}(\Omega)$ are the following forward Fourier transforms of the quantities $B_{P,SV,SH,BB,DD}$ given in Chapter 5:

$$\hat{B}_{P,SV,SH,BB,DD}(\Omega) = \int_{0}^{\infty} d\xi \, B_{P,SV,SH,BB,DD}(\xi) e^{i\xi\Omega}, \qquad (7.5)$$

where Ω denotes one of the quantities $2\kappa_a, 2\kappa_b, \kappa_-, \kappa_+$ or $-\kappa_-$.

Let us, like in the previous chapter, assume that all autocorrelation as well as all crosscorrelation functions of the velocity and density fluctuations are exponential (with an identical correlation distance l but different variances σ). By performing the inverse Fourier transformation (7.1) and using results (6.7) in equation (5.35), we obtain again formula (7.2) with the functions $S_{1SH}(t), S_{1P,SV}(t), S_{2P,SV}(t)$, and $S_{3P,SV}(t)$ written now explicitly as:

$$S_{jP,SV,SH}(t) = \exp\left(-\Lambda_{jP,SV,SH} - \tau_{jP,SV,SH}\right)\Upsilon_{jP,SV,SH} \times$$

$$\times \left[\delta(\tau_{jP,SV,SH}) + \Lambda_{jP,SV,SH} \frac{I_1(2\sqrt{\Lambda_{jP,SV,SH}\tau_{jP,SV,SH}})}{\sqrt{\Lambda_{jP,SV,SH}\tau_{jP,SV,SH}}}\right], \qquad (7.6)$$

if $\tau_{1SH}, \tau_{jP,SV,SH} > 0$, and $S_{jP,SV,SH}(t) = 0$ if $\tau_{1SH}, \tau_{jP,SV,SH} < 0$. Here I_1 is the modified Bessel function of order 1; index j is equal to 1, 2 or 3 for the P- and

SV-waves and it is equal to 1 for the SH-wave. Also the following abbreviations are used in equations (7.6):

$$\tau_{j P,SV,SH} = \Lambda_{j P,SV,SH} + \frac{t}{l\Omega_{j P,SV,SH}},$$

$$\Lambda_{j P,SV,SH} = \frac{L\beta_{j P,SV,SH}}{l\Omega^2_{j P,SV,SH}}, \quad \Upsilon_{j P,SV,SH} = \frac{1}{l|\Omega_{j P,SV}|},$$

$$\beta_{1 P,SV,SH} = B_{P,SV,SH}(0), \quad \beta_{2P} = \beta_{2SV} = B_{BB}(0), \qquad (7.7)$$

$$\beta_{3P} = \beta_{3SV} = B_{DD}(0), \quad \Omega_{1P} = 2a^{-1}X, \quad \Omega_{1SV,SH} = 2b^{-1}Y,$$

$$\Omega_{2P} = -\Omega_{2SV} = b^{-1}Y - a^{-1}X, \quad \Omega_{3P} = \Omega_{3SV} = a^{-1}X + b^{-1}Y.$$

Obviously, equations (7.2) and (7.6) are not much more convenient for numerical calculations than the initial Fourier transform (7.1). However, some general facts can be observed from them. For instance, in an exponential medium the transient transmissivity (7.2) consists of the sum of two parts: (i) a scaled δ-pulse (coherent part) and (ii) the pulse with the Bessel functions and their convolutions (incoherent part). The coherent part is produced by direct (multiple-free) transmission. The incoherent part is produced by (short peg-leg) multiples of arbitrary order. The shape of the P-wave transient transmissivity does not significantly differ from that of scalar waves (i.e. the SH-wave or pressure-wave transient transmissivity). It should be noted here that for P- and SH-waves the transient transmissivity is minimum delay (because the real and imaginary parts of the logarithm of the time-harmonic transmissivity satisfy the Kramers-Krönig relationship). It is in agreement with the theorem of minimum delay of stratigraphic filtering proved for vertically incident waves by Robinson and Treitel (1980, pp. 321-325).

The minimum-delay property does not hold for the SV-SV transmissivity. For oblique incidence the incoherent part of the SV-wave transmissivity arrives earlier (on account of internal mode conversion) than the coherent part (which has no mode conversion). For the P- and SH-wave transmissivity the coherent part constitutes always the first arrival. The coherent part diminishes exponentially with increasing L, whereas the incoherent part diminishes much slower. In fact for thick stacks of layers it is only the incoherent part that can be observed.

Figure 25 shows three pulse signatures of the transient transmissivities for the P-, SV-, and SH-waves. For these illustrations the same medium parameters are used as in the previous chapter. These signatures were computed by the numerical Fourier transform using equation (7.1). The traveltimes of the transient transmissivities were reduced in order to see all three signatures in a non-overlapping form on the same plot and in equal time and amplitude scales. Due to discretization it is not possible to observe the coherent part (i.e. the exponentially scaled δ - pulse) of the transmissivities. However, an abrupt onset of the signals for P- and SH-waves can be seen. This is not the case for the SV-wave.

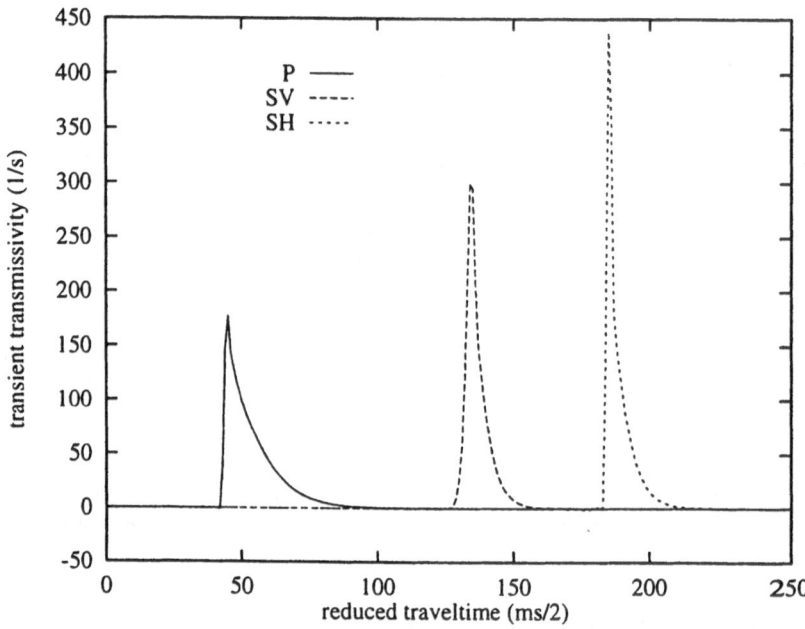

Fig. 25. Signatures of the transient transmissivities of P-, SV-, and SH- waves. The medium parameters are the same as in Figure 12. The thickness $L = 500m$. The angle of incidence $\vartheta = 30°$. (From Shapiro and Hubral, 1996).

A simple approximation of the transient transmissivity can be obtained for media (not only exponential ones), which provide for high frequencies a larger attenuation coefficient than for low ones. In such media (for a sufficiently large L) only low-frequency components of the transient transmissivity survive. This means that in equation (7.5) we can use the Taylor expansion of the exponential factor and restrict ourselves to the first term (which is equal to 1) only. This approximation yields the *Gaussian pulse*:

$$\tilde{T}_{P,SV,SH}(t) = \frac{1}{2\sqrt{\pi L \tilde{\gamma}_{P,SV,SH}^{low}}} \exp\left[\frac{-(t - px - \tilde{\psi}_{P,SV,SH}^{low} L)^2}{4L \tilde{\gamma}_{P,SV,SH}^{low}}\right], \tag{7.8}$$

where

$$\tilde{\gamma}_P^{low} = \int_0^\infty d\xi \left[B_P(\xi) + B_{BB}(\xi) + B_{DD}(\xi)\right],$$

$$\tilde{\gamma}_{SV}^{low} = \int_0^\infty d\xi \left[B_{SV}(\xi) + B_{BB}(\xi) + B_{DD}(\xi)\right], \qquad (7.9)$$

$$\tilde{\gamma}_{SH}^{low} = \int_0^\infty d\xi B_{SH}(\xi).$$

A similar approximation (expressed in other terms and called 'diffusion approximation') was given by Burridge and Chang (1989).

A comparison of the pulse (7.8) with the result (7.2) is presented in Figure 26 for the P-wave. Again, we have used for the computations the same medium parameters as in the previous chapter. Here we show the P-wave transient transmissivities at the point ($x = L \tan \vartheta, z = L$) for $L = 1200m$ (pulses to the left) and $L = 6000m$ (pulses to the right) for the angle of incidence $\vartheta = 30°$. The traveltime is reduced by 95 % of the corresponding traveltimes for vertical incidence in the reference medium. For the transient transmissivities in the case of $L = 1200m$ and $L = 6000m$ the origin of the time axes has been set equal to $0.95 \times 1200m/4000\frac{m}{s} = 0.285s$ and $0.95 \times 6000m/4000\frac{m}{s} = 1.425s$, respectively. Obviously, for large values of L the approximation (7.8) yields satisfactory results.

It is important to note that this approximation provides a systematic error in the time of first arrivals because of its non-causal nature. For example, in an exponential medium the first arrivals of the P-wave transient transmissivity is defined by the high-frequency limit of the phase velocity. The peaks of the amplitude arrive later. However, the approximation (7.8) neglects the high-frequency components of the transmissivity.

7.2 Numerical Transmissivity and Pulse Stabilization

Next we present four numerical simulations involving the transient transmissivity (7.1). In Figure 27 we display the transient P-wave transmissivity as a function of the travel time for three angles of incidence: $0°, 30°$ and $45°$. The transient transmissivity is received at the point ($x = L \tan \vartheta, z = L$). The layered medium

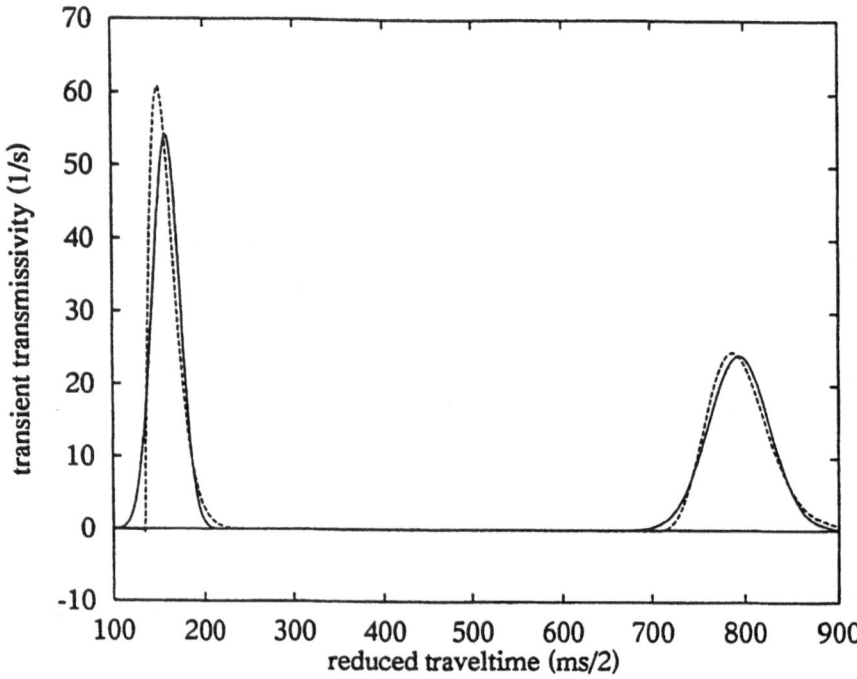

Fig. 26. P-wave transient transmissivity computed by formula (7.1) (dashed line) and by the approximation (7.8) (solid line) for $L = 1200m$ (leftmost pulses) and $L = 6000m$ (rightmost pulses) for the angle of incidence $\vartheta = 30°$. The medium parameters are the same as for Figure 12. The traveltime is reduced by 95 per cent of the traveltimes for vertical incidence. (From Shapiro and Hubral, 1996).

is now $1000m$ thick. Such thickness or even a larger one is no problem, as the exact transient transmissivity can be constructed without any phase-unwrapping. The fast Fourier transform does it automatically but causes another problem: the roughness of the analytical curves numerically computed by formula (7.1). In spite of these numerical effects we notice once again a good agreement between the theoretical and numerical results. An interesting phenomenon can be observed, namely an increase of the peak amplitude of the transient transmissivity with the angle of incidence. Though at first glance surprising this is, nevertheless, in perfect agreement with the decrease of the attenuation for increasing angles of incidence in the range of low frequencies (Figure 13). In the range of high frequencies we can state that the larger the angle of incidence the higher is the attenuation. This explains why the coherent part (the spike at the time close to 255ms in Figure 27) of the transient transmissivity can be observed for vertical incidence only.

The corresponding functions are plotted for $L = 2000$m in Figure 28. In order to make this plot we reduced the traveltimes by 95 percent of the traveltime in the case of a vertically incident P-wave in a homogeneous reference medium with

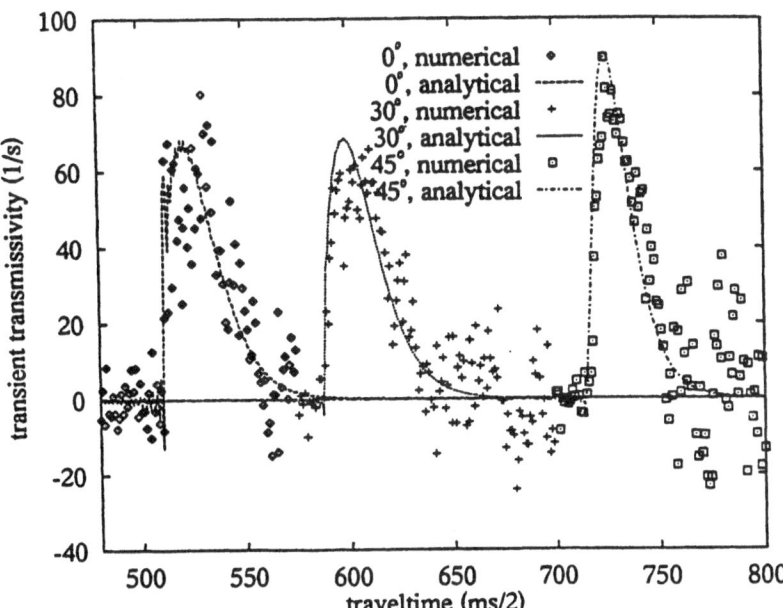

Fig. 27. Comparison of the numerical and theoretical estimations of the P-wave transient transmissivity (superimposed in three separated windows) for three angles of incidence. The parameters of the medium are the same as for Figure 12. $L = 1000m$. (From Shapiro and Hubral, 1996).

thickness 2000m. A comparison between Figure 27 and Figure 28 shows that there are less fluctuations in the numerically computed transmissivities for the larger medium. This is a *consequence of the self-averaging phenomenon*, which is also called sometimes the *pulse stabilization* (see, e.g., Lewicki, 1994, and Lewicki et al. 1994).

Figures 29 and 30 present the same comparison for $L = 1000m$ and $L = 2000m$ respectively between numerical and theoretical results for the transient SV-wave transmissivity. Again a good agreement between numerical and analytical results is observed.

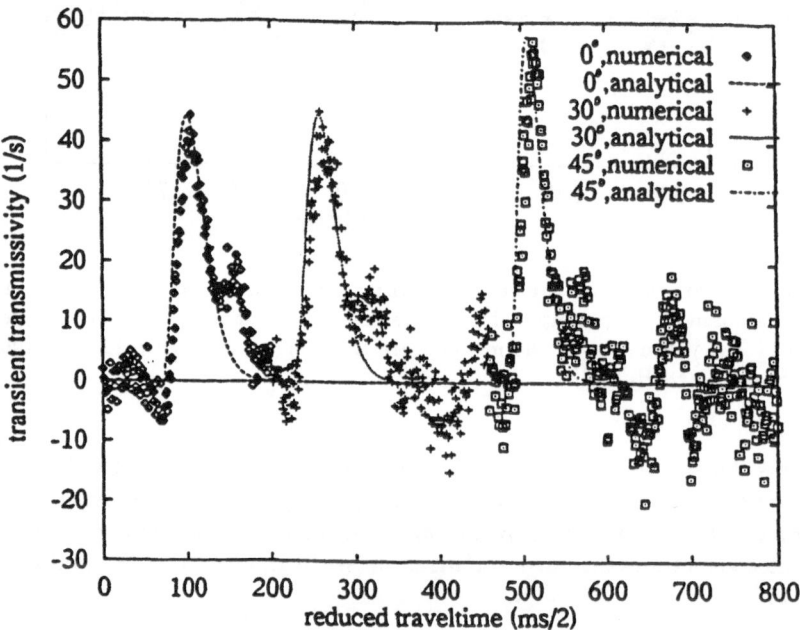

Fig. 28. The same as Figure 27 but for $L = 2000m$. (From Shapiro and Hubral, 1996).

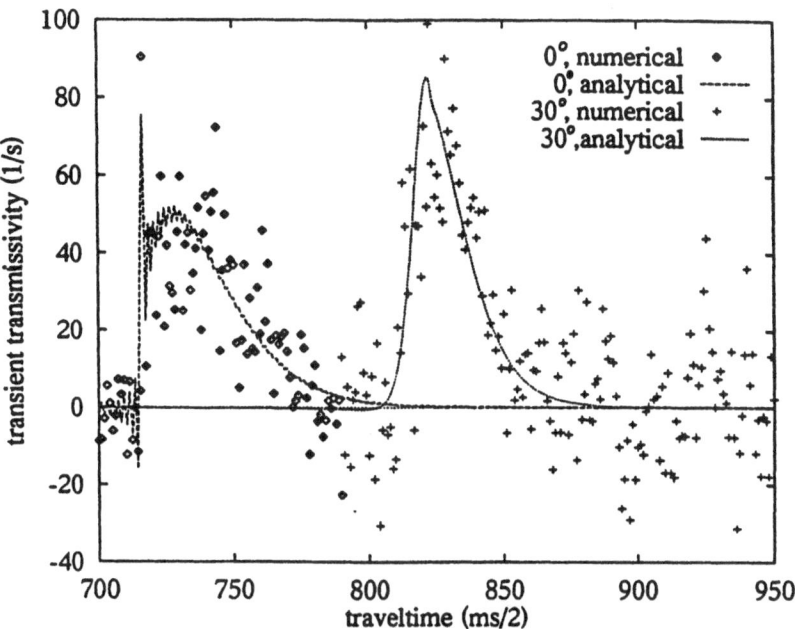

Fig. 29. Comparison of the numerical and theoretical estimations of the SV-wave transient transmissivity (superimposed in two separated windows) for two angles of incidence. The parameters of the medium are the same as for Figure 12. $L = 1000m$.(From Shapiro and Hubral, 1996).

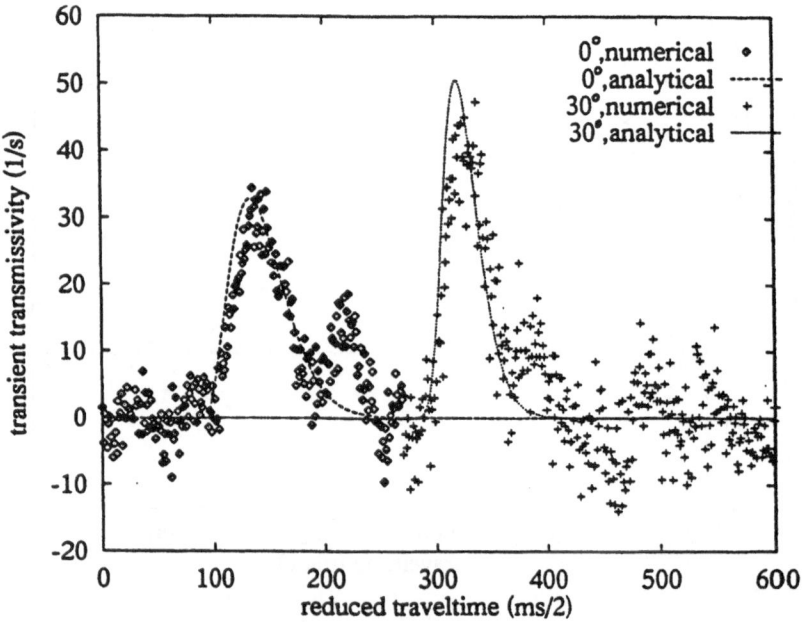

Fig. 30. The same as Figure 29 but for $L = 2000m$. (From Shapiro and Hubral, 1996).

8 Stratigraphic Filtering and Amplitude Variation with Offset

The elastic P-wave reflection response of a target (e.g., one single or more interfaces of a layered stack) below a multilayered overburden is subjected to two-way stratigraphic filtering. Under two-way stratigraphic filtering we understand all effects of the multilayered overburden, which make the reflection responses of a target different from reflection responses of this target overlain by a homogeneous overburden. In the first approximation (consistent with ODA-formulas, see e.g., Lewicki and Burridge 1996) the effect of two-way stratigraphic filtering can be described as a product of two transmissivities: from the source to the target reflector and from the reflector to the receiver. As we have seen in the previous chapters the transmission of seismic waves is significantly affected by stratigraphic filtering in multilayered overburdens. Therefore, reflection imaging and *amplitude-variation-with-offset (AVO) analysis* necessarily have to consider this effect. The generalized ODA formulas (4.19), (5.36), (5.37) can be used to correct the offset-dependent reflection response for the effects of two-way stratigraphic filtering. An application of these analytical formulas needs a statistical macro model of the overburden, which includes the standard deviations and correlation lengths of the elastic parameters and density fluctuations.

Such a correction procedure for the AVO-analysis method was developed by Widmaier et al. (1995) and Widmaier et al. (1996). Later the effect of thin-layering on seismic amplitudes was further investigated by Widmaier (1996) using well-log data collected in thick sediment formations of the North Sea (the log data were provided by STATOIL). In this chapter we follow the lines of Widmaier, (1996) however, simplifying and shortening his treatment.

It is important to note here, that all generalized ODA formulas have been derived for incident plane waves. However, in seismic practice we always work with concentrated sources of elastic energy. Therefore, strictly speaking, a correction method for spherical waves should be developed. Numerical simulations and practical experience show, however, that the plane-wave based correction works very well.

The AVO-correction procedure, which we describe in this chapter, can be applied for any type of waves incident on and reflected from a target. In practice, however, the AVO analysis is usually applied for the incident and reflected P-waves only. For this reason we consider in the following elastic P-waves mainly.

8.1   Correction Method

In the previous chapters we expressed the equivalent-dynamic-medium time-harmonic transmissivity of multilayered structures (see Figure 1) in the following form:

$$T_p \propto e^{\left(i(\psi_p L + \omega p x - \omega t) - \gamma_p L\right)} \tag{8.1}$$

We recall here the most important notations: the parameter L describes the thickness of the overburden and p denotes the horizontal slowness. The quantities ψ_p and γ_p are the vertical phase increment and the attenuation coefficient of the plane P-wave transmitted through the overburden, respectively. They both can be obtained from an angle-dependent combination of the auto- and crosscorrelation functions of P-wave-velocity, S-wave-velocity and density logs (equation (5.36)).

Equation (8.1) shows that the equivalent-dynamic-medium transmissivity is decreased like $exp(-\gamma_p L)$. Additionally, the transmissivity is characterized by a phase shift. Thus, an inverse filter to the stratigraphic one must be applied. To correct for the amplitude effect of the stratigraphic filtering the transmissivity must be multiplied by $exp(\gamma_p L)$ for the down- and upgoing ray path. The correction of the phase is also possible. However, usually the phase correction is not necessary to recover the amplitude-versus-offset behavior of the target.

To demonstrate the proposed AVO correction procedure, we consider the elastic model shown in Figure 31: an isolated target reflector underlying a layered overburden. Additionally on this figure a Common-Mid-Point (CMP) system of shots and receivers is shown. The thickness of the layered stack is 1000m. The velocity v_0 of the homogeneous parts of the medium below and above the layered stack is chosen equal to the normal-incidence velocity determined by ray theory (i.e., v_0 is determined by the averaged slowness $v_0 = <1/v(z)>^{-1}$).

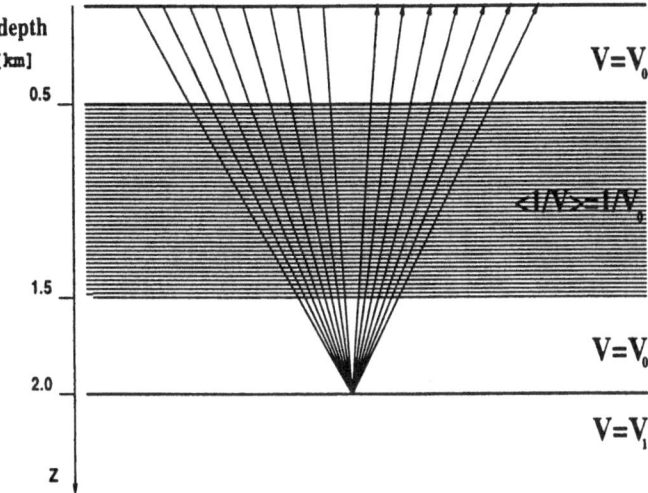

Fig. 31. *A Common-Mid-Point observational system and the geometry of the elastic multilayered model. (This Figure has been taken, with kind permission, from Widmaier, 1996, p. 32.)*

Elastic P-wave model (L =1000m)	v_p	v_s	ϱ
mean value < x >	3000	1730	2.2
standard deviation σ_{xx} [%]	8.0	8.0	0.0
correlation length l_{xx} [m]	9	9	–

Table 1. *Statistical parameters for the model shown in Figure 32.*

Figure 32 shows the P-wave velocity, S-wave velocity and density logs that correspond to the chosen model. Mean velocities, relative standard deviations and correlation lengths of P- and S-wave velocity fluctuations are listed in Table 1. The density was kept constant. The exponentially correlated P- and S-wave velocity fluctuations were generated with a uniform correlation length. Both synthetic velocity logs were taken to be completely correlated by assuming a constant v_p/v_s ratio.

A numerical synthetic-seismogram simulation was performed to compute a 2-D Common Shot Record (CSR). Note that for the geometry shown in Figure 31 the CSR and CMP records are equal. The CSR record is shown in Figure 33. One can observe the common-shot reflection response of the target horizon at around $1.3s$ zero-offset traveltime. For shot-receiver offsets higher than 2 km wavefield interference occurs between the target's P-wave reflection and P to S converted reflections from the layered overburden.

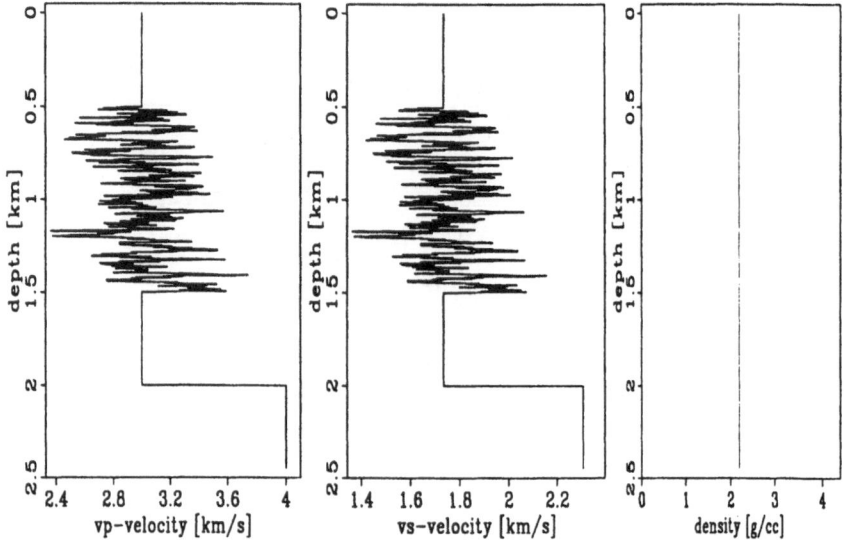

Fig. 32. v_p, v_s and density logs of the multilayered model.(This Figure has been taken, with kind permission, from Widmaier, 1996, p. 33.)

The common-shot data were copied for laterally displaced shot positions and the traces were sorted to *common-offset gathers*. These were depth migrated with an $2\frac{1}{2} - D$ *true-amplitude common-offset Kirchhoff migration* scheme (Widmaier et al, 1996). In Figure 34 the *resulting AVO curve* (obtained from the primary-reflection image of the target reflector) is compared with the expected angle-dependent reflection coefficient in the case of a homogeneous overburden. As the true-amplitude migration scheme, based on ray theory, ignores thin-layer effects the obtained amplitudes are obviously too low. The undulations of the amplitudes at larger offsets are due to interference between the target's P-wave response and P to S converted reflections from the overburden as described above (see Figure 33).

In order to take stratigraphic filtering effects into account, a usual (ray-theory based) processing scheme has to be completed by using the generalized ODA formula for elastic P-waves. In such a way the attenuation of the downgoing P-wave transmissivity from a source to a target and the attenuation of the upgoing transmissivity from the target to a receiver can be corrected. For this, in addition to a traditionally-used macro-velocity model, a statistical macro model for the entire multi-layered overburden is required. In our particular model there are, because of the chosen medium characteristics (exponential correlation function of the medium fluctuation, constant density, homogeneous background, constant v_p/v_s ratio and uniform correlation length), only 4 parameters, which constitute the statistical macro model. These are the correlation length l, the variances of

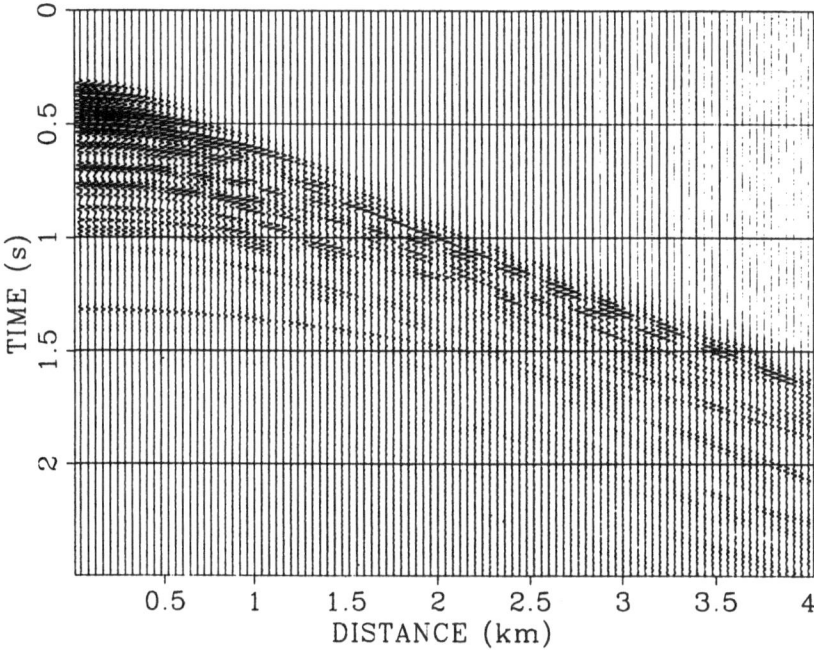

Fig. 33. *Common-Shot Vertical-Displacement Record for the elastic multilayered model shown in Figure 31. (This Figure has been taken, with kind permission, from Widmaier, 1996, p. 33.)*

P- and S-wave-velocities logs, and their covariance.

The AVO correction has been performed as follows. For every offset the corresponding initial angle for the ray, which is incident on the multi-layered stack, was calculated. Then, the frequency- and angle-dependent two-way attenuation factor $exp(-2\gamma_p L)$ for the up- and downgoing wave, or, in terms of equation (8.1), the absolute value of the equivalent-dynamic-medium two-way transmissivity is calculated using the statistical macro model. The inverse two-way attenuation factor, computed for the dominant frequency and for the corresponding offset, has then been multiplied with the reflection coefficients obtained after true-amplitude Kirchhoff migration. The resulting AVO curve after application of this multi-layer correction is shown in Figure 35. The correction leads to a significantly improved AVO behavior (compare with Figure 34).

Wapenaar et al, (1994) proposed a similar approach to the AVO correction for the effects of stratigraphic filtering. Their correction method was restricted to acoustic media only and also based on the true-amplitude-migration processing scheme. In the following sections we will demonstrate another (even more simplified) way of the correction.

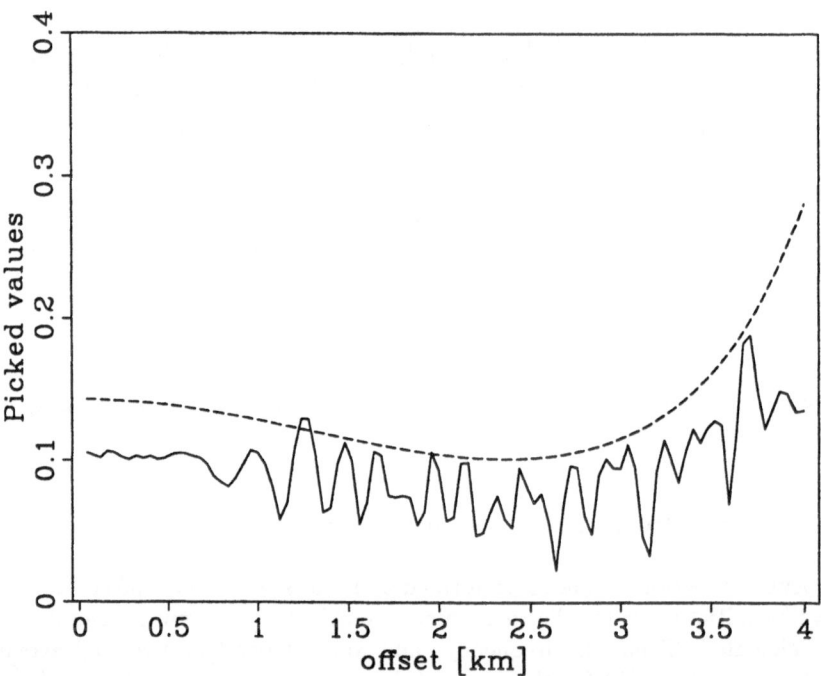

Fig. 34. *Comparison of the measured AVO curve (solid line) with the reflection co-efficients computed for the homogeneous overburden (dashed line). The offset range corresponds to angles of incidence between 0 and 45 degrees. (This Figure has been taken, with kind permission, from Widmaier, 1996, p. 36.)*

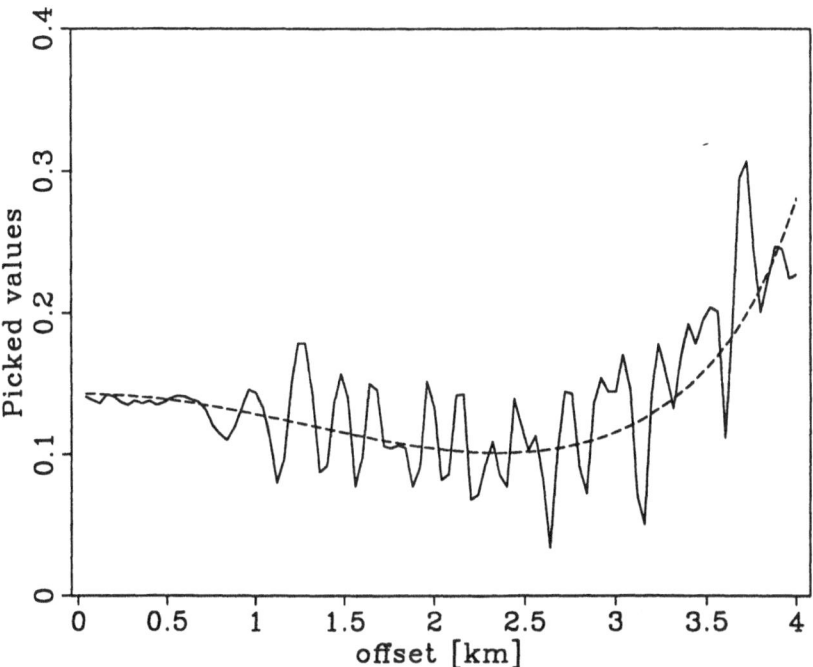

Fig. 35. *Comparison of the AVO curve after application of the multi-layer correction (solid line) with the reflection coefficient for the homogeneous overburden (dashed line). (This Figure has been taken, with kind permission, from Widmaier, 1996, p. 38.)*

8.2   Well-Log Data

A modeling study similar to the previous one was performed on a real-log data set (Widmaier, 1996). Logs for a well located in the North Sea were generously provided by STATOIL. For this log P-wave velocities, S-wave-velocities and densities were recorded in the depth range between 750m and 2970m (true vertical depth). Two depth intervals have been chosen and isolated from that well data for the following consideration. The first interval corresponds to the *Hordaland stratigraphic group* and is located between 948m and 1755m depth. The second one corresponds to the *Shetland stratigraphic group*. The top of the Shetland group is at depth level 2011m with the bottom at 3006m. Recording was only done down to 2970m. In the following these two depth intervals are referred to as the *Hordaland Model* and the *Shetland Model*.

Applying the coarse blocking described as follows, Widmaier (1996) removed effects of non-stationarity. In reality the P-wave velocity, S-wave velocity, and density values $(v_p(z), v_s(z), \varrho(z))$ were measured along the Hordaland and Shetland group with a depth increment of 0.1 m. Both, the Hordaland model as well as the Shetland model have been blocked into approximately 20 layers within a 1000m depth interval. A comparison of the original P-wave velocity logs, S-wave velocity logs, and density logs with the correspondingly blocked versions is shown in Figure 36 and Figure 37 for the Hordaland and the Shetland model, respectively.

For the study of the amplitude effects of stratigraphic filtering for the Hordaland and the Shetland models, their corresponding original and blocked log data intervals were separated from the complete well log. The depth locations of the log interval were shifted upwards. Each formation was embedded between two homogeneous half spaces. For all models an artificial single target horizon was introduced below the thin layers to investigate the effect of stratigraphic filtering on the reflector's AVO response (Table 2). For computing synthetic seismograms seismic source and receivers were placed in the uppermost layer for all models. The source is positioned below the coordinate origin at 100m depth. Recording of the vertical component of the displacement velocity was simulated. The receiver spacing was 20 m. A non-free surface was chosen to avoid ghosts and multiple reflections caused by the surface.

Figures 38 and 39 show corresponding plots of the artificially created thinly-layered and blocked log models for the Hordaland and the Shetland group. For each considered geological formation two different shot records were calculated. The first one was based on the blocked log interval of the formation. The well log data with the original sample increment of 0.1m were used as modeling input for the second shot record. In other words, for the Hordaland as well as for the

Artificial Target	P-wave velocity m/s	S-wave velocity m/s	density g/cc	V_p/V_s
Hordaland Model				
above target	2135	674	2.13	3.17
below target	1900	850	1.90	2.38
Shetland Model				
above target	3463	1834	2.54	1.89
below target	3200	2000	2.30	1.60

Table 2. *Parameters used for the artificial isolated target horizons in the Hordaland and Shetland models.*

Shetland group the artificial model consists of more than 8000 fully elastic thin layers.

A comparison of the numerical-simulation results for the blocked and thinly-layered models is shown in Figures 40 and 41 for the Hordaland and Shetland groups, respectively. The backscattered energy from the thin layers and the re-flection response from the target horizon can clearly be identified at approximately 1.3 s (Hordaland model) and 0.9 s (Shetland model) zero-offset two-way time.

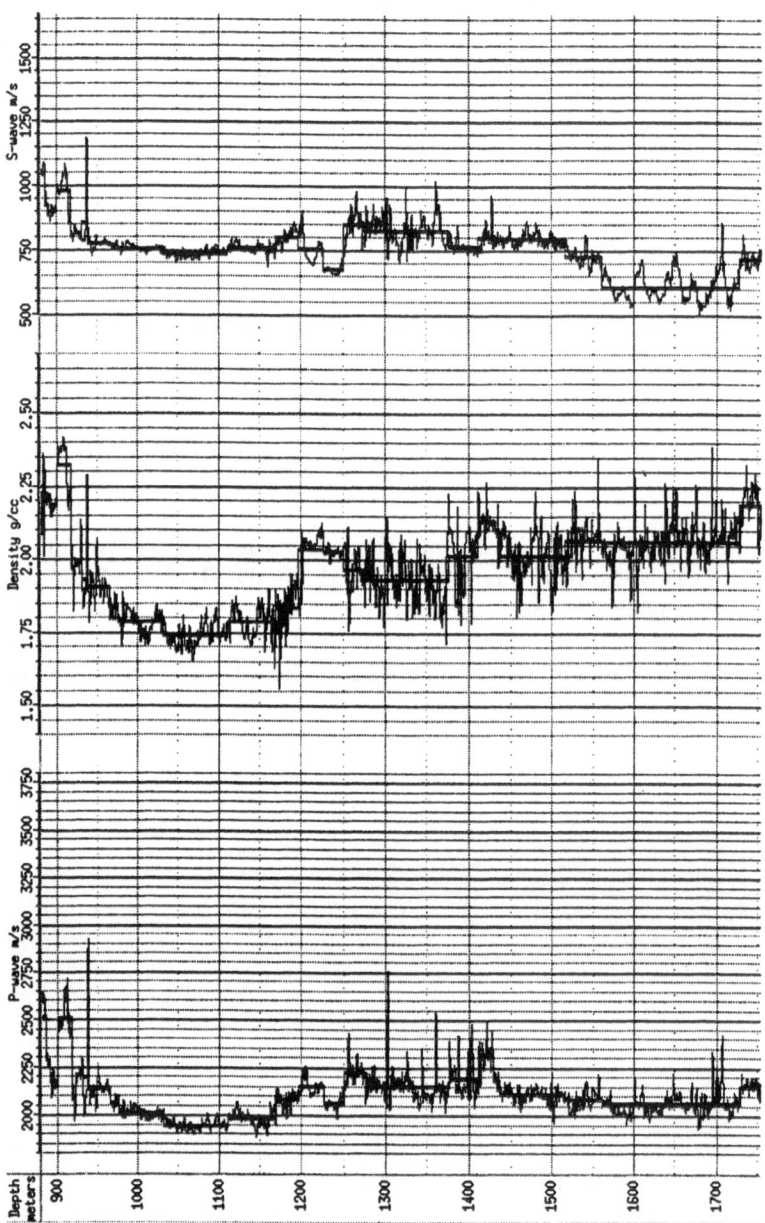

Fig. 36. Hordaland model: Comparison of the original v_P, ϱ and v_s logs with the corresponding blocked logs. Depth range of the Hordaland group is from 948m (top) to 1755m (bottom). (This Figure has been taken, with kind permission, from Widmaier, 1996, p. 51.)

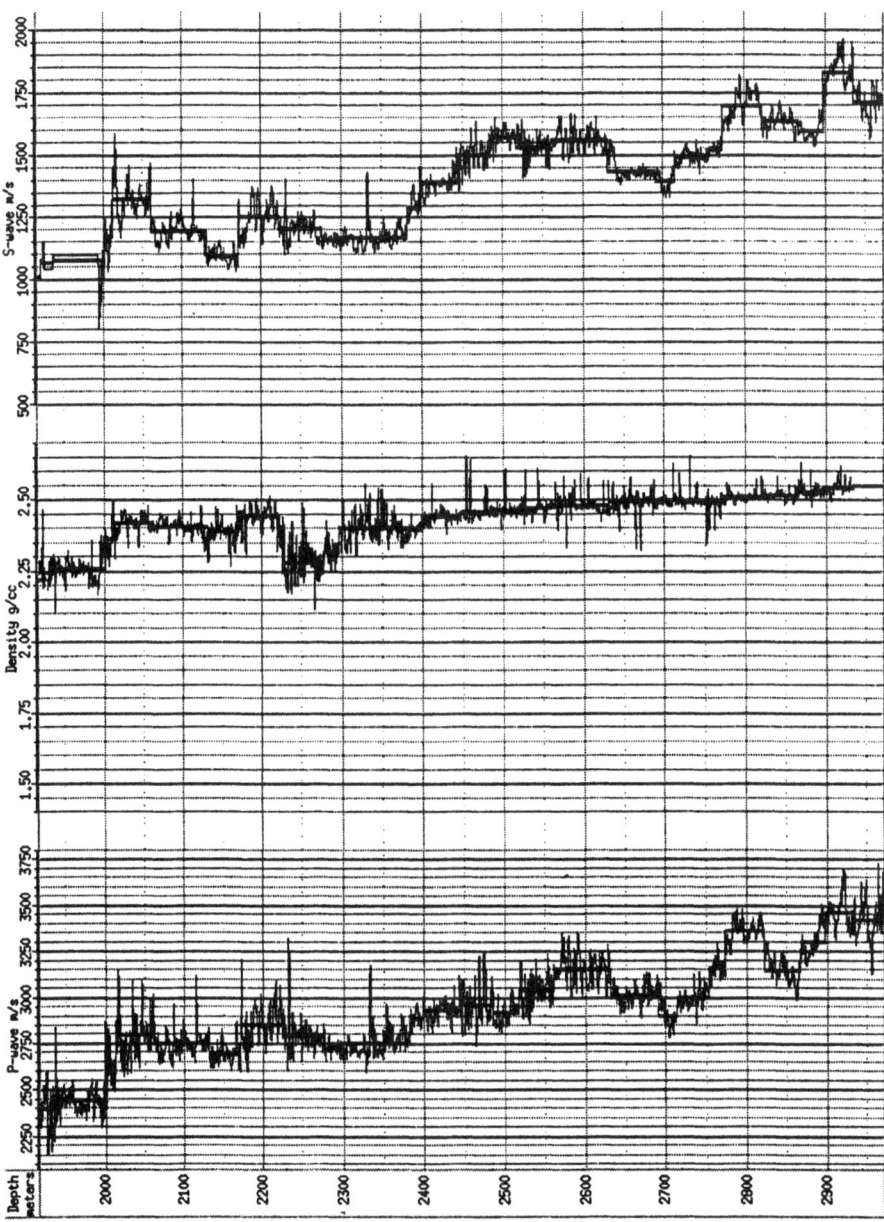

Fig. 37. *Shetland model: Comparison of the original v_p, ϱ and v_s logs with the corresponding blocked logs. Depth range of the Shetland group is from 2011m (top) to 2970m (bottom). (This Figure has been taken, with kind permission, from Widmaier, 1996, p. 52.)*

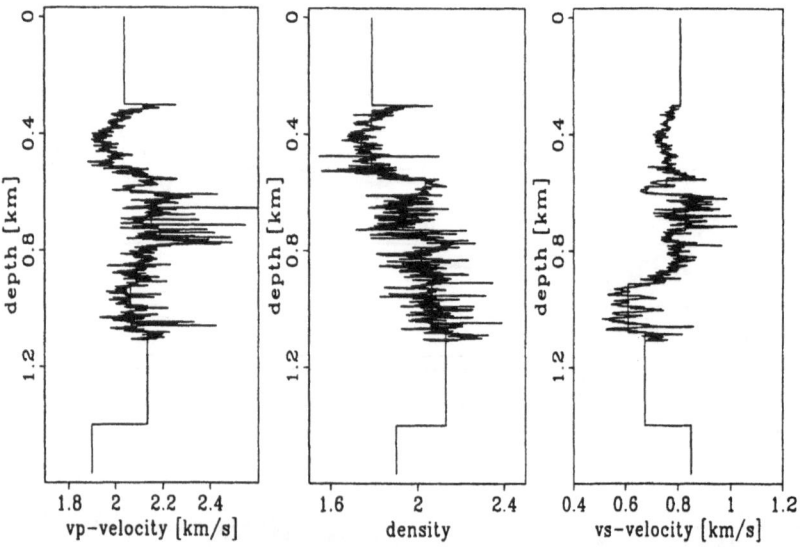

Fig. 38. *Hordaland model: The depth-shifted original and the blocked log intervals. The homogeneous parts below and above the stratigraphic group and the target horizon were introduced artificially. (This Figure has been taken, with kind permission, from Widmaier, 1996, p. 54.)*

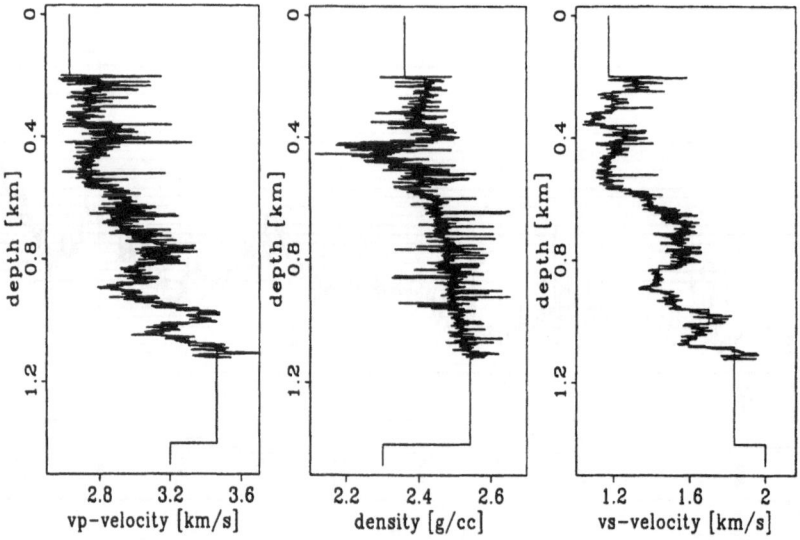

Fig. 39. *Shetland model: The depth-shifted original and the blocked log intervals. The homogeneous parts below and above the stratigraphic group and the target horizon were introduced artificially. (This Figure has been taken, with kind permission, from Widmaier, 1996, p. 54.)*

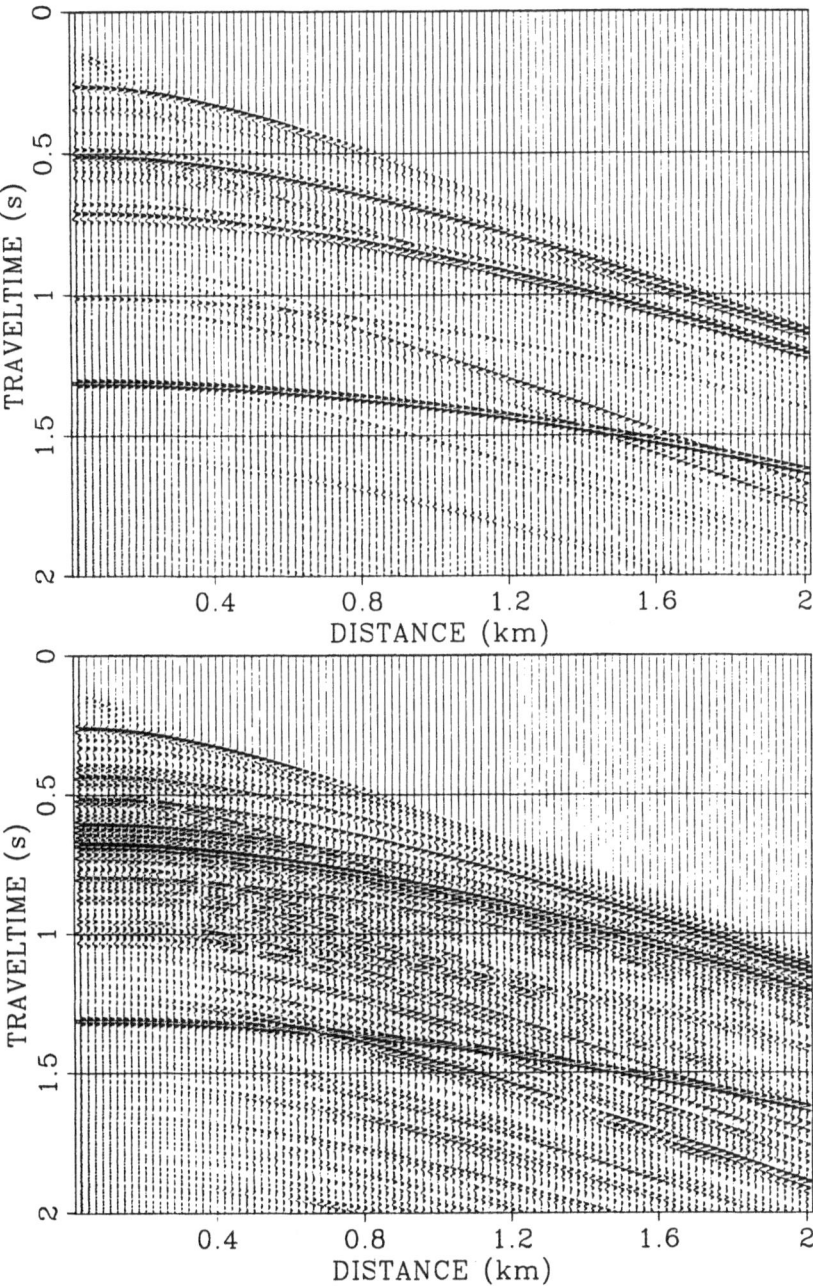

Fig. 40. *Hordaland Model: Comparison of the seismograms, calculated for the blocked version of the model and the thinly-layered version of the model. The target reflection event can be identified as the isolated event around 1.3 s zero-offset two-way traveltime. (This Figure has been taken, with kind permission, from Widmaier, 1996, p. 56.)*

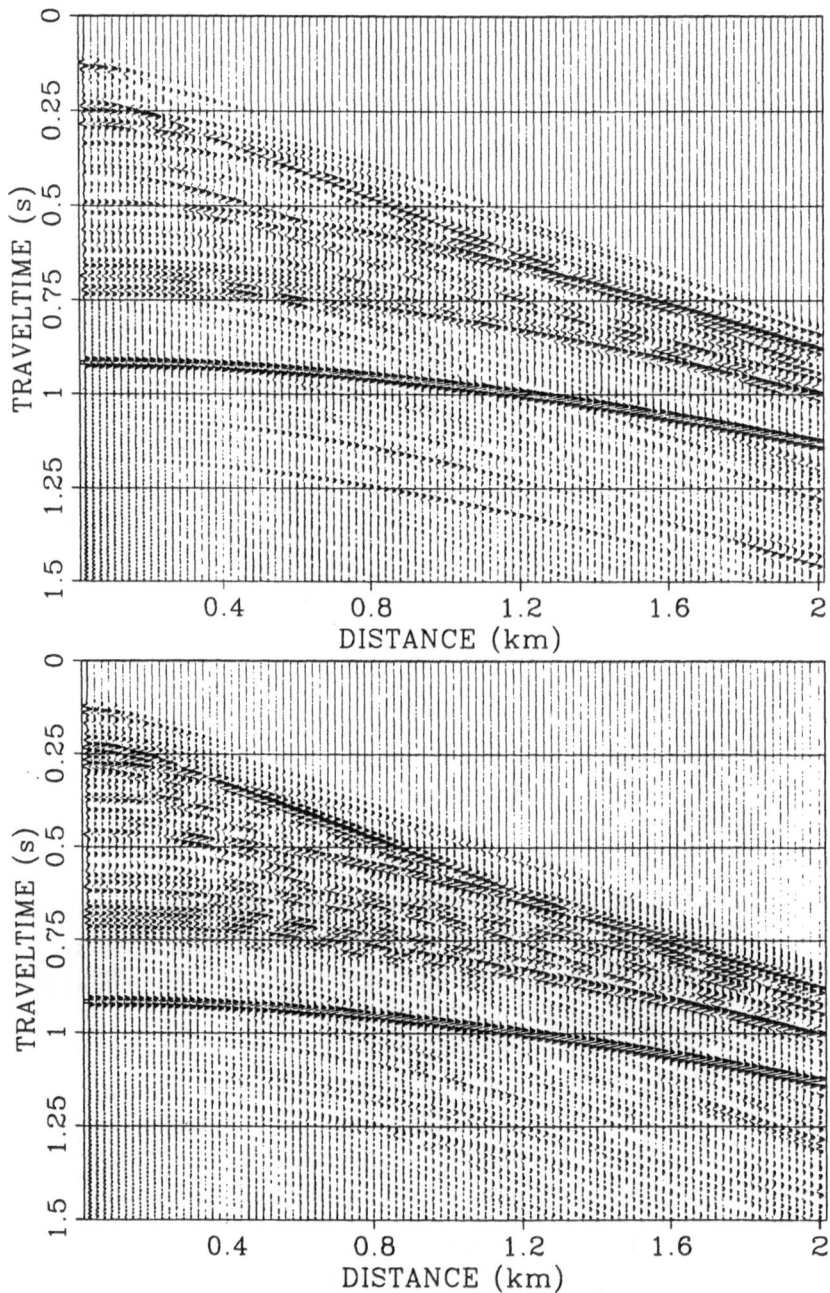

Fig. 41. *Shetland Model: Comparison of the seismograms calculated for the blocked version of the model and the thinly-layered model. The target reflection event can be identified as the isolated event around 0.9 s zero-offset two-way traveltime. (This Figure has been taken, with kind permission, from Widmaier, 1996, p. 57.)*

8.3 Amplitude Processing

Widmaier (1996) showed that in addition to the true-amplitude Kirchhoff-migration-based processing scheme (see Section 8.1) the AVO effect of the stratigraphic filtering can be corrected also in another way. He applied to the Hordaland and Shetland models a conventional AVO processing of CMP gathers, as recommended, e.g., in Castagna (1993). This true amplitude processing was then completed by a multi-layer correction based on the generalized ODA formula.

Due to the 1-D geometry of the created geological models it was possible to handle the synthetic common-shot records like CMP records. The following processing steps were applied to each of the four common-shot records shown in the previous section before correcting them for the effects of stratigraphic filtering:

- *Spherical divergence correction:* The geometrical spreading loss was corrected using normal-moveout (NMO) velocities (or RMS-velocities) and zero-offset traveltimes (Ursin, 1990, formula 23).
- *Emergence-angle correction:* By assuming straight rays to the target horizon and using the corresponding NMO-velocity the ray parameter was estimated. As the ray parameter is constant along the ray, the emergence angle can be determined if the velocity of the first layer is known.
- *NMO correction:* Based on the velocities of the blocked log data the NMO correction was introduced.
- *F-K filtering:* It can be observed in Figure 40 and Figure 41 that for larger offsets interference occurs between the target's P-wave reflection and the P to S converted reflections from the overburden. The interfering events could be separated by transforming the NMO-corrected data into the F-K domain. To prevent similar distortions as seen in the AVO response of Figure 34, the waves converted in the overburden were removed by a standard fan filter.

Therefore, no migration has been applied to the data obtained for Hordaland and Shetland models. After the processing described above the peak values of the target reflections were collected for both generated models. Then a constant scaling factor was applied to fit the AVO curves obtained from numerical simulations to the near-offset range of the true offset-dependent reflection coefficients (exact Zöppritz equation for parameters given in Table 2) of the target horizons.

This scaling was done for practical reasons. Neither seismic data acquisition (e.g., unknown source strength, receiver coupling) nor seismic processing can provide and guarantee amplitudes in an absolute and true-amplitude sense. Usually, sufficient accuracy and amplitude preservation for the AVO analysis can only be given in the sense of a relative offset dependence of the reflection coefficient.

Finally, the AVO curves of the blocked and thinly-layered models were plotted together with the true offset-dependent reflection coefficients. For the Hordaland model the offset-dependence of the reflection response of the target is not affected significantly by the propagation through the coarsely blocked model (Figure 42, top). Except for a slight distortion for offsets around 1500m (probably due to interference), the AVO curve and the exact reflection coefficients coincide. However, no coincidence can be observed for the thinly-layered model of the Hordaland group (Figure 42, bottom). The picked values show a different offset behavior. Obviously, the values could not be fitted to the curve of the true reflection coefficients by scaling with a constant factor. The amplitudes for the larger offsets remain too low compared to the expected ones. Therefore, the seismic amplitudes must have been attenuated significantly by scattering in the thinly-layered overburden.

Similar results can be observed for the Shetland model. Again, the blocked version of the overburden has no significant influence on the measured AVO response of the target horizon (Figure 43, top). For the original Shetland model the amplitudes again show attenuation effects for larger offsets (Figure 43, bottom). However, the influence of the thinly-layered overburden of the Shetland model is less pronounced than that in the case of the Hordaland model.

The modeling results above show that the transmission losses caused by the few introduced interfaces of the blocked models have no significant influence on the offset behavior of the AVO curves.

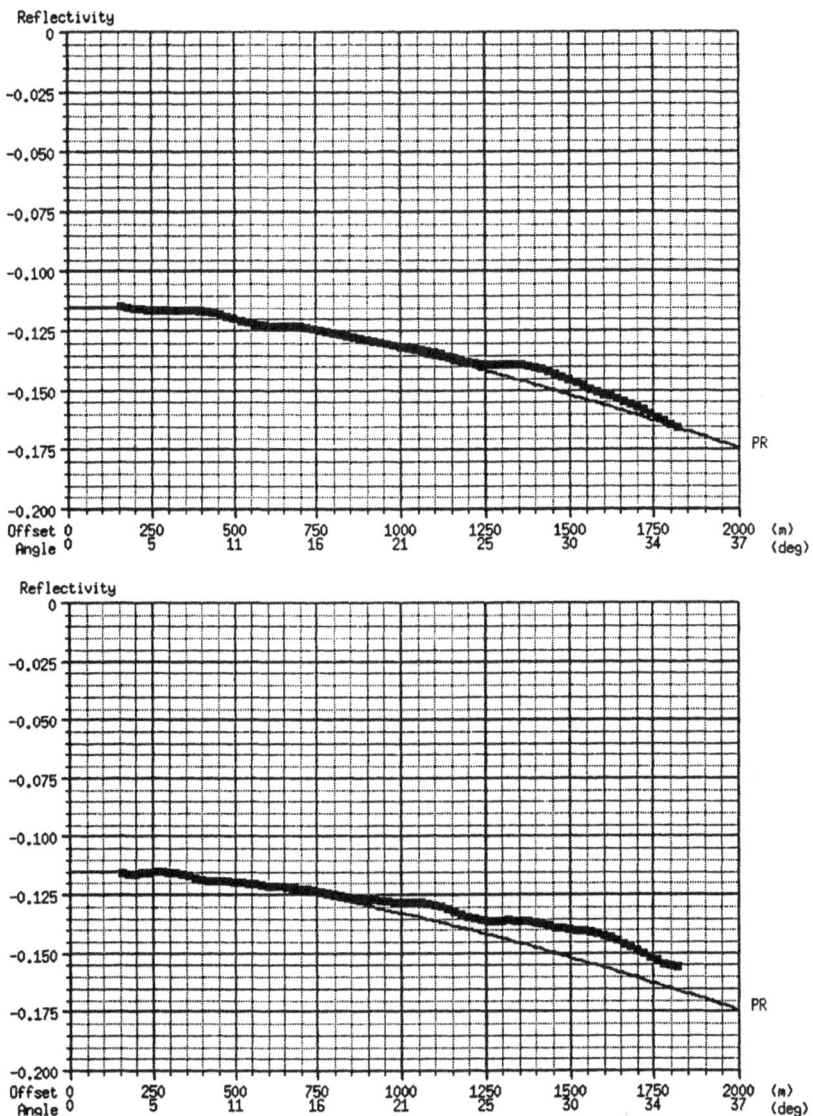

Fig. 42. *AVO curve for the blocked Hordaland model (top) and the thinly-layered Hordaland model (bottom) compared with the true reflection coefficients (thin curves). The AVO curves were normalized by constant scale factors to fit the true reflection coefficients in the near-offset range. (This Figure has been taken, with kind permission, from Widmaier, 1996, p. 61.)*

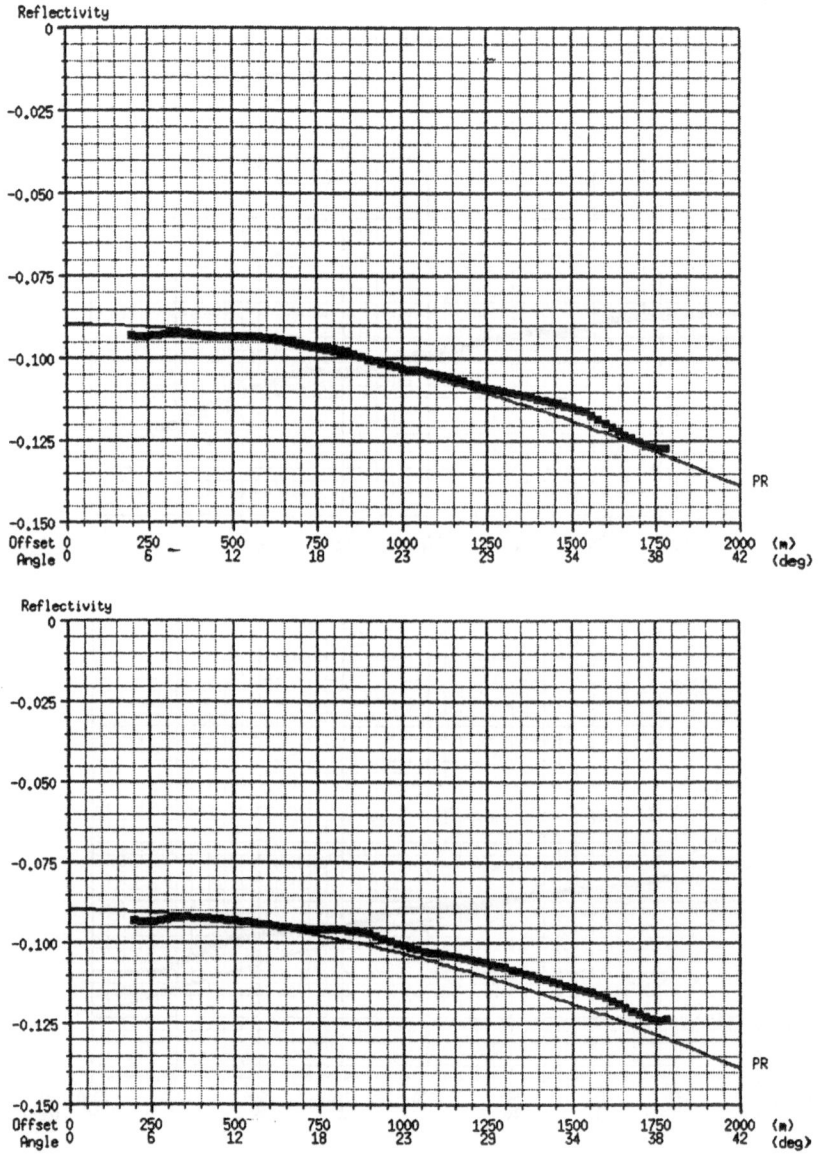

Fig. 43. *AVO curve for the blocked Shetland model (top) and the thinly-layered Shetland model (bottom) compared with the true reflection coefficients (thin curves). The AVO curves were normalized by constant scale factors to fit the true reflection coefficients in the near-offset range. (This Figure has been taken, with kind permission, from Widmaier, 1996, p. 62.)*

8.4 Estimation of Statistical Macro Models from Well Logs

In order to apply the generalized-ODA-formula based correction to AVO curves to account for the stratigraphic filtering statistical macro models of overburdens are necessary. Such models describe the auto- and crosscorrelations of the small-scale medium fluctuations. Under small scale we understand here a scale, which is much smaller than the target's depth.

Usually, however, a practical problem is that measured well log data have a non-stationary appearance. For instance, the physical parameters of the Hordaland and Shetland models do not fluctuate around constant averages in contrast to some synthetic examples of this book. The background values vary slightly corresponding to lithological changes. This background trend has to be estimated and separated from the fluctuating parts for the statistical analysis of the small-scale variations. Then, statistical parameters can be estimated by correlating and cross-correlating these isolated fluctuations. Thus, the statistical analysis of the Hordaland and Shetland group can be divided into the following steps:

- *Trend removal*: According to the observations from the AVO modeling results (Figure 42 and 43) and the quantitative estimation of the two-way transmission loss, the offset-dependent amplitude response of the target horizon is not affected significantly by the coarsely blocked models. All offset-dependent influence seems to be attributed to the fluctuations. Thus, the blocked log data can be considered as the background trend and can be subtracted from the corresponding original log data. The obtained residual fluctuations for v_p, v_s and ϱ of both formations are shown in Figure 44 and Figure 45.

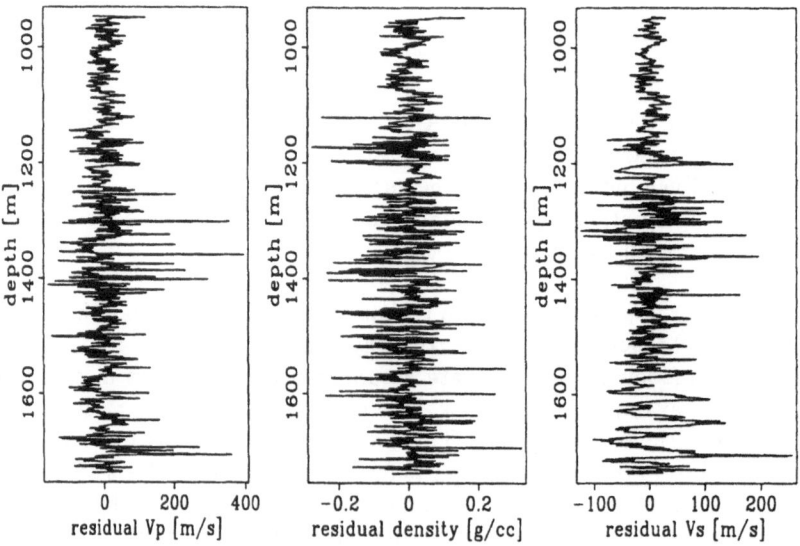

Fig. 44. *Residual* v_p, v_s *and density fluctuations of the Hordaland model after subtraction of the corresponding blocked log data. (This Figure has been taken, with kind permission, from Widmaier, 1996, p. 64.)*

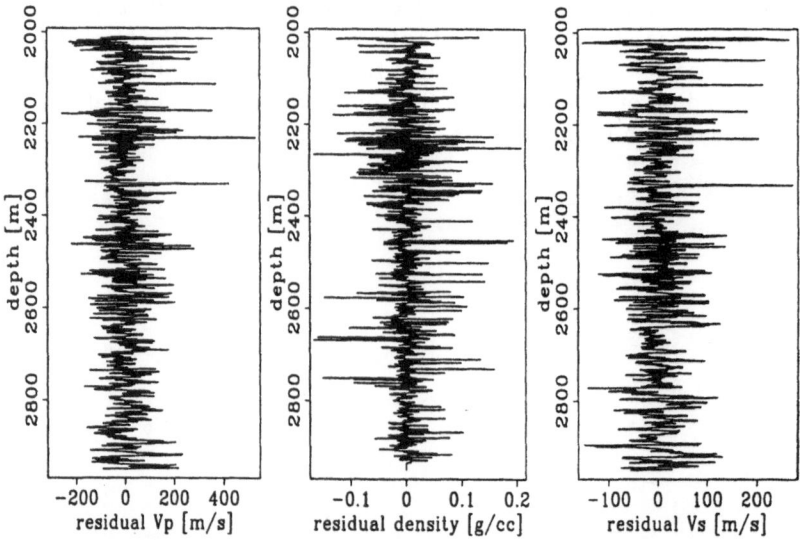

Fig. 45. *Residual* v_p, v_s *and density fluctuations of the Shetland model after subtraction of the corresponding blocked log data. (This Figure has been taken, with kind permission, from Widmaier, 1996, p. 64.)*

- *Correlation*: It is assumed that the considered intervals of the North Sea
well log are characterized by exponential correlation functions. Therefore,
the auto- and cross-correlations of the isolated fluctuations have to decay like
$exp(-\xi/l)$. Here, ξ denotes the correlation lag and l the correlation length.
Auto- and crosscorrelations were calculated for the residual fluctuations of
v_p, v_s, and ϱ residuals. By a least-square algorithm an exponential function $\sigma^2 exp(-\xi/l)$ was fitted to each correlation. The least-square fitting was
restricted to small correlation lags where the correlation has only positive
values. In Figure 46 and 47 the correlation functions of the residual random
fluctuations are compared to the fitted exponential functions. All autocorrelation functions are normalized by the square of the standard deviation σ.
The crosscorrelations are normalized by the product of the standard deviations of the two considered fluctuating logs. Therefore, the zero-lag value of
the cross-correlation equals to the correlation coefficient R_{xy}:

$$R_{xy} = \frac{\sigma^2_{xy}}{\sigma_{xx}\sigma_{yy}}. \tag{8.2}$$

The variables σ_{xx} and σ_{yy} denote the standard deviation of x and y, σ^2_{xy}
defines the covariance of x and y. The comparison of the correlation results
with the fitted curves shows that the choice of an exponential function seems
to be appropriate. The positive value range is well approximated by these
functions for small lags. Due to the very low crosscorrelation between the
residuals of the P-wave velocity and the density and between the S-wave
velocity and the density in the Shetland model (Figure 47) fitting was not
possible. The correlation coefficient R_{xy} was set equal to zero.

In the case that the assumption of an exponential correlation would have
failed for the log data, one could try to use for the fit other correlation
functions. For instance, a von Karman function or a linear combination of
exponential functions is useful for characterizing a more general case of frac-
tal-type stratifications.

- *Parameter extraction:* All estimated parameters for the fluctuations of both
models are listed in Table 3 and 4. The correlation and the crosscorrelation
lengths are defined by the parameter l, which characterizes the decay of
the fitted exponential function. The normalized standard deviations of the
residual fluctuations are obtained after normalizing the standard deviations
by the mean values of the corresponding original $v_p(z)$, $v_s(z)$ and $\varrho(z)$ log
intervals. This normalization and especially the definition of the mean values
by a simple averaging of the logs seems to be a very rough approximation.
However, related tests with synthetic gradient models (Knoth et al., 1996,

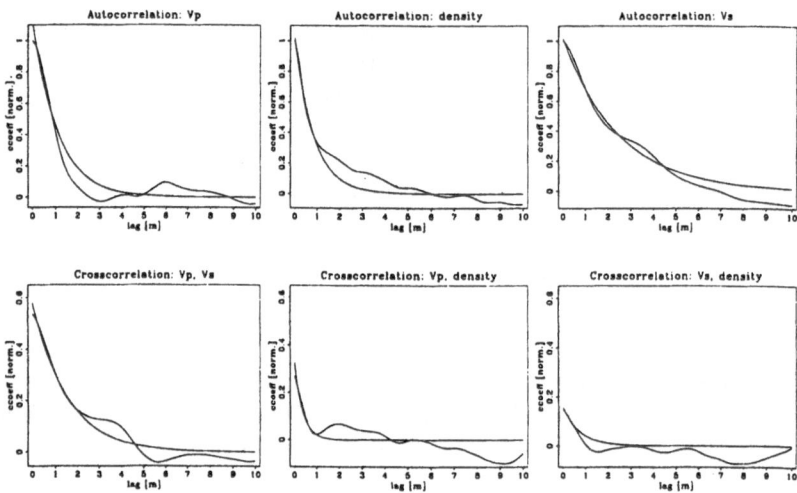

Fig. 46. *Normalized auto- (first row) and crosscorrelations (second row) of the residual fluctuations of the Hordaland model with fitted exponential functions. (This Figure has been taken, with kind permission, from Widmaier, 1996, p. 66.)*

Knoth 1996) have shown that the generalized ODA formula is not sensitive with respect to small variations of the mean values with depth.

It is obvious that the normalized standard deviations calculated for the residual v_p, v_s, and ϱ fluctuations of the Shetland model are smaller than the one for the shallower Hordaland model (Table 3). This fact was already observed qualitatively in the modeling results, where the effect of the stratigraphic filtering on the AVO response was smaller for the Shetland model. In general, a higher normalized fluctuation for the S-wave velocities than for the P-wave velocities was obtained. Correlation lengths were estimated for both log intervals in the order of one or two meters.

The crosscorrelation between P-wave and S-wave velocities has a significant impact on the angle dependence of scattering attenuation and transverse isotropy in general. Crosscorrelation coefficients of about 0.54 (Hordaland) and 0.63 (Shetland) were estimated from the log data (Table 4). This is a strong evidence that S-wave velocity logs are needed for exact amplitude studies. Introduction of a pseudo-shear wave log based on lithological information in terms of constant v_p/v_s or constant Poisson's ratio, when real data are not available, would falsely simulate a perfect crosscorrelation ($R_{v_p v_s} = 1$) and, therefore, underestimate the angle-dependent scattering and other related effects.

Hordaland (L =800m)	v_p	v_s	ϱ	Shetland (L =960m)	v_p	v_s	ϱ
mean value $< x >$	2090	740	1.9	mean value $< x >$	2990	1420	2.4
norm. standard deviation σ_x [%]	3.0	5.6	3.6	norm. standard deviation σ_x [%]	2.7	3.4	1.3
autocorrelation length l_x [m]	1.1	2.5	0.9	autocorrelation length l_x [m]	1.2	1.6	0.6

Table 3. *Parameters of statistical macro models calculated from the autocorrelations of the residual fluctuations for the Hordaland and the Shetland groups.*

Hordaland (L =800m)	v_p, v_s	v_p, ϱ	v_s, ϱ	Shetland (L =960m)	v_p, v_s	v_p, ϱ	v_s, ϱ
cross-correlat. coeffic. R_{xy}	0.54	0.27	0.15	cross-correlat. coeffic. R_{xy}	0.63	~0	~0
cross-correlat. length l_{xy} [m]	1.5	—	0.36	0.74	cross-correlat. length l_{xy} [m]	0.82	–	–

Table 4. *Parameters of statistical macro models calculated from the crosscorrelations of the residual fluctuations for the Hordaland and the Shetland groups.*

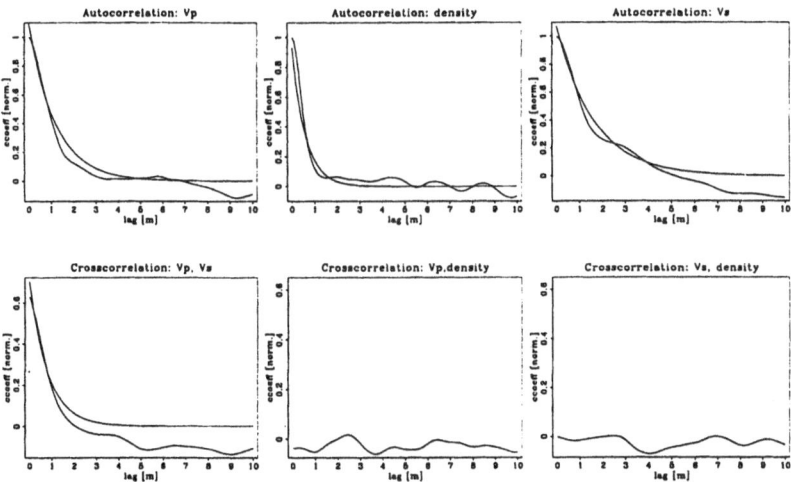

Fig. 47. *Normalized auto- (first row) and crosscorrelations (second row) of the residual fluctuations of the Shetland model with fitted exponential functions. (This Figure has been taken, with kind permission, from Widmaier, 1996, p. 66.)*

8.5   Correction of AVO Responses

After the derivation of macro-statistical parameters the dynamic-equivalent-medium attenuation coefficient γ can be estimated by the generalized ODA formula. Using only the dynamic part of this formula the angle- and frequency-dependent transmission loss due to the thin layers is described by

$$T = e^{-\gamma L} , \qquad (8.3)$$

where L is the thickness of the considered formation. Here, γ is the attenuation coefficient in its most general form for exponentially correlated fluctuations. It is defined by the 12 statistical parameters in Tables 3 and 4.

The stratigraphic-filtering distorted AVO response of the Hordaland model (Figure 42) was corrected successfully (Figure 48, bottom) by multiplying it with the offset-dependent factor T_0^2/T^2, where T_0 is the quantity T at zero offset. The correction factors were computed for the main frequency of the wavelet. The corrected AVO response now exhibits the same offset dependence as the true reflection coefficients of the target. A comparison of the normalized AVO responses for the blocked Hordaland model and its thinly-layered versions (uncorrected and corrected) emphasizes that the generalized ODA formula enables us to take the residual fluctuations into account. It removes the misfit between the blocked macro-model and the original one (Figure 48, top). Application of the correction procedure to the less distorted AVO response of the Shetland model (Figure 43) leads to a similar improvement (Figure 49).

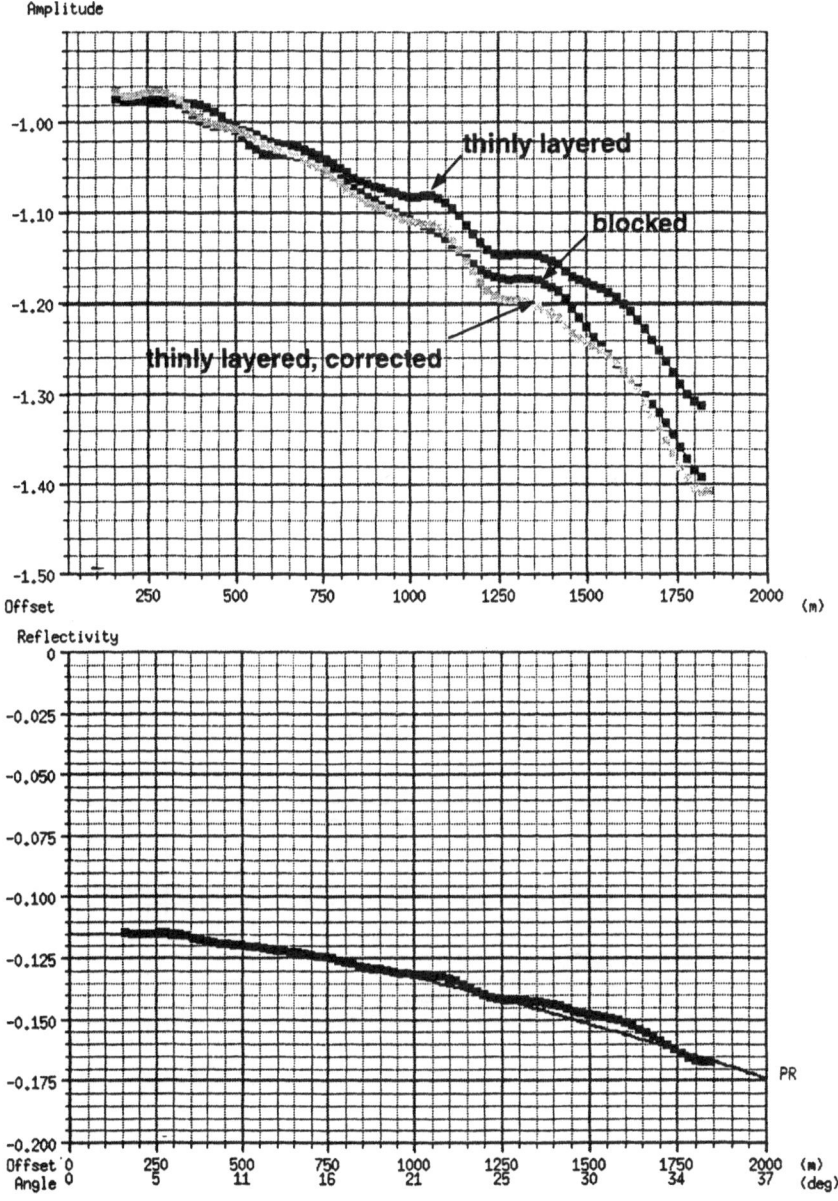

Fig. 48. Top: *Comparison of normalized AVO responses for the thinly-layered Hordaland model before (black) and after correction (grey) with the response of the blocked model.* **Bottom:** *Corrected AVO response for the thinly-layered Hordaland model compared with true reflection coefficients (thin). (This Figure has been taken, with kind permission, from Widmaier, 1996, p. 72.)*

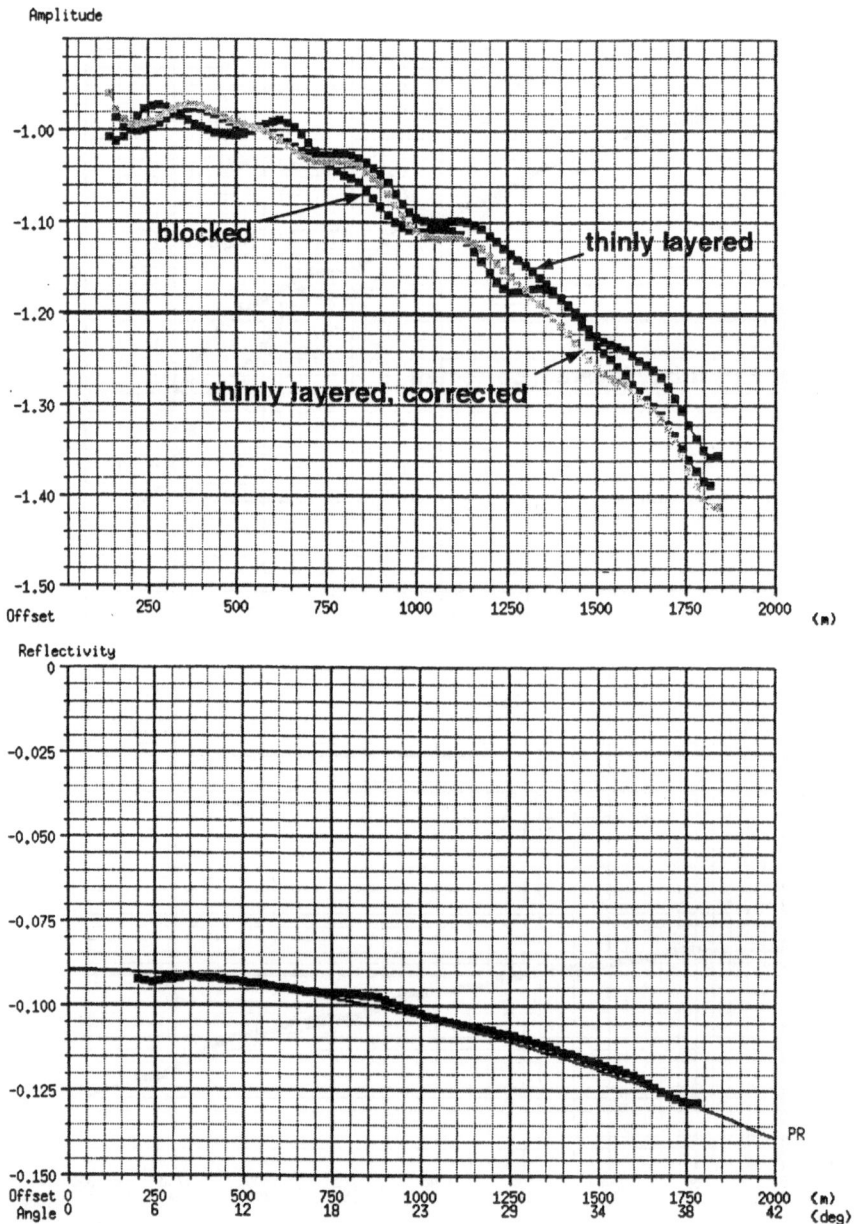

Fig. 49. Top: *Comparison of normalized AVO responses for the thinly-layered Shetland model before (black) and after correction (grey) with the response of the blocked model.* **Bottom:** *Corrected AVO response for the thinly-layered Shetland model compared with true reflection coefficients (thin). (This Figure has been taken, with kind permission, from Widmaier, 1996, p. 73.)*

8.6   Scattering or Intrinsic Absorption?

Whether simple Q compensation or a conventional decibel-per-second gain function would sufficiently account for the stratigraphic-filtering effects in AVO processing schemes, is a question, which still remains open. In AVO modeling the overburden is usually replaced by an inelastic macro model, for which constant values of the quality factor Q are chosen. In some North-Sea case study (see, e.g., Wrolstad, 1993) all attenuation effects in the overburden are included in a macroscopic constant Q. This *effective quality factor* was assumed to be 200 for P-waves and introduced in the following way:

$$T_{int} = e^{-\gamma_{int}\frac{2L}{\cos\vartheta}}, \quad \gamma_{int} = \frac{\pi f}{Q_{int}c_0} \quad . \tag{8.4}$$

Here, Q_{int} denotes the frequency- and angle-independent quality factor for the corresponding intrinsic attenuation coefficient γ_{int}. The amplitude loss depends on the length of the raypath only (or on the traveltime). The term $2L/\cos\vartheta$ accounts for the length of the downgoing and upgoing raypath for an arbitrary angle of incidence ϑ.

For comparison of both effects, scattering and absorption, in the work of Widmaier (1996) the intrinsic-attenuation coefficient was chosen to be equivalent to the scattering-attenuation coefficient γ_{TL} of the Hordaland group for vertical incidence. Figure 50 shows clearly the difference in the angle-dependence of the transmission behavior between scattering attenuation of the Hordaland group and the constant-Q absorption. Due to the chosen magnitude of intrinsic attenuation, coincidence is observed for small offsets. However, the transmission loss caused by the thin layering at larger offsets is higher for all frequencies.

When looking at Figure 50 one should always have in mind that the magnitude of scattering attenuation for normal incidence is in general lower than the total attenuation measured in zero-offset VSP's. Following the estimates given in Schoenberger and Levin (1974), the attenuation due to thin layering accounts for 30 to 50 % of the total frequency-dependent absorption (normal incidence). The relation of these values to the actual lithology is obvious.

In connection with this it is interesting to note, that the inversion of the scattering attenuation for the statistical macro model of the overburden made by Zien (1993) and discussed in Section 6.6 shows excellent agreement with the sonic-log observations (see Figure 24). This seems to be in contradiction with the previous paragraph, because Zien attributed the complete frequency-dependent part of the attenuation to scattering (interpreting a significant frequency-independent part of the attenuation coefficient as the effect of not-compensated geometrical spreading and non-ideal receiver coupling). Thus, it is possible, that in the future, careful estimations of the attenuation will change our ideas on the relationship between the scattering and inelastic absorption.

For the Hordaland group the Q_{sc} value would be around 1900 (main frequency). It is important to note here that the Hordaland and Shetland models are characterized by very small standard deviations of the medium fluctuation. The fluctuations can be much stronger. For instance, the geological formation, reflection-coefficient series of which is shown in Figure 5, is characterized by the standard deviation of the acoustic impedance of 0.14. This leads to the scattering attenuation of one order larger than those for the Hordaland model. In the case considered by Zien (1993) the standard deviation of the acoustic impedance was about 0.1. Thus, the scattering attenuation was approximately 5 times larger than those for the Hordaland model. However, even such small-fluctuating models as the Hordaland and the Shetland ones show the significance of the stratigraphic-filtering effects and the importance of corresponding AVO corrections.

Indeed, a synthetic seismogram was computed based on an inelastic version of the Hordaland model (Figure 38). For simplicity, a homogeneous distribution of the quality factor, where Q_{int} equals 200 in all layers, was assumed. Exactly the same processing steps as described in Section 8.3 were applied now to the inelastic data.

Figure 51 shows the picked, normalized AVO curve. Due to the additional absorption the amplitude distortion is larger than in the corresponding purely layered model. After application of standard Q-compensation (using professional processing software), where the quality factor first was set equal to 200, the compensated AVO curve (Figure 52) still shows a different offset behavior. Using the Q value of about 180, taking now into account also the scattering attenuation at zero offset (we call this approximation pseudo-effective), still does not lead to a correct improvement of the AVO behavior at the larger offsets (Figure 53).

With regard to the considered model, where inelasticity was combined with thin-layering effects, a combined correction procedure seems to be most natural. Sequential compensation of absorption (assuming that the intrinsic $Q_{int}=200$ is known) and application of the generalized ODA formula for thin-layer correction (based on statistical parameters given in Tables 3 and 4) provides the best AVO recovery (Figure 54).

$$\frac{1}{Q_{eff}(\vartheta)} = \frac{1}{Q_{sc}(\vartheta)} + \frac{1}{Q_{int}} \tag{8.5}$$

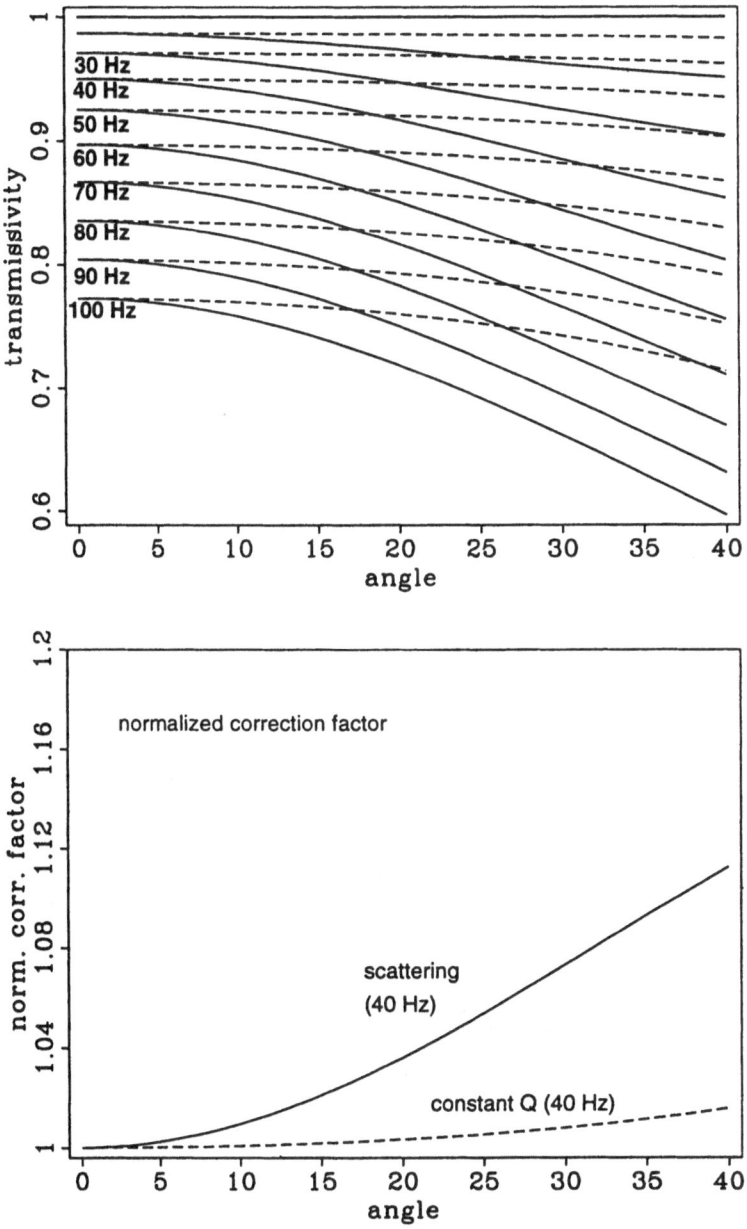

Fig. 50. *Angle dependencies of the absolute values of the transmissivity (top) and normalized correction factors (bottom, main frequency) for a medium with scattering only (solid curves) and for a medium with a constant-Q intrinsic absorption only (dashed curves). Absorption was chosen to be equal to scattering attenuation of the Hordaland model for each frequency at normal incidence. (This Figure has been taken, with kind permission, from Widmaier, 1996, p. 75.)*

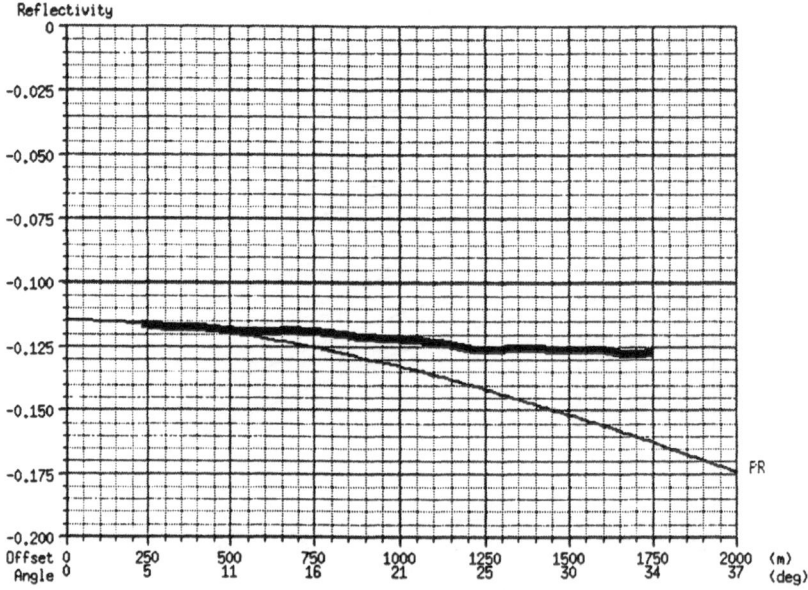

Fig. 51. *AVO response for the inelastic* $(Q_{int} = 200)$, *thinly-layered Hordaland model. (This Figure has been taken, with kind permission, from Widmaier, 1996, p. 78.)*

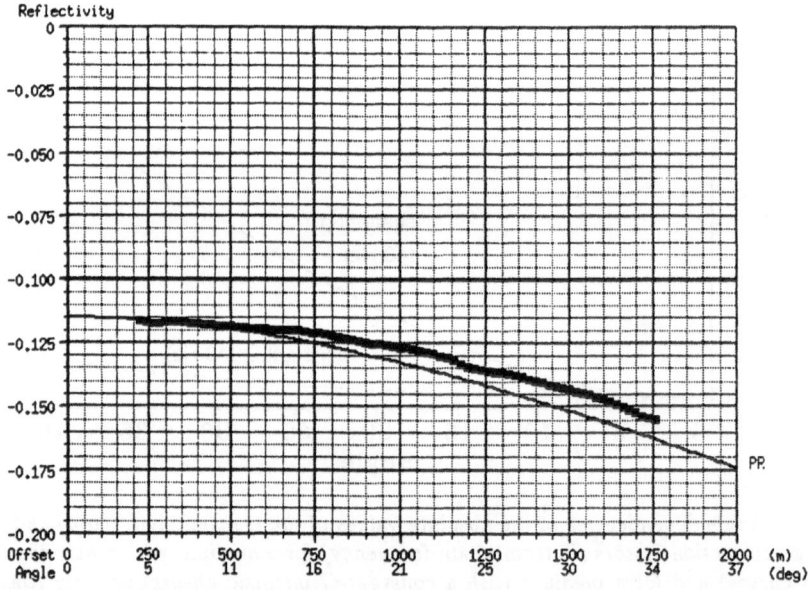

Fig. 52. *AVO response after Q-compensation* $(Q_{int} = 200)$. *(This Figure has been taken, with kind permission, from Widmaier, 1996, p. 78.)*

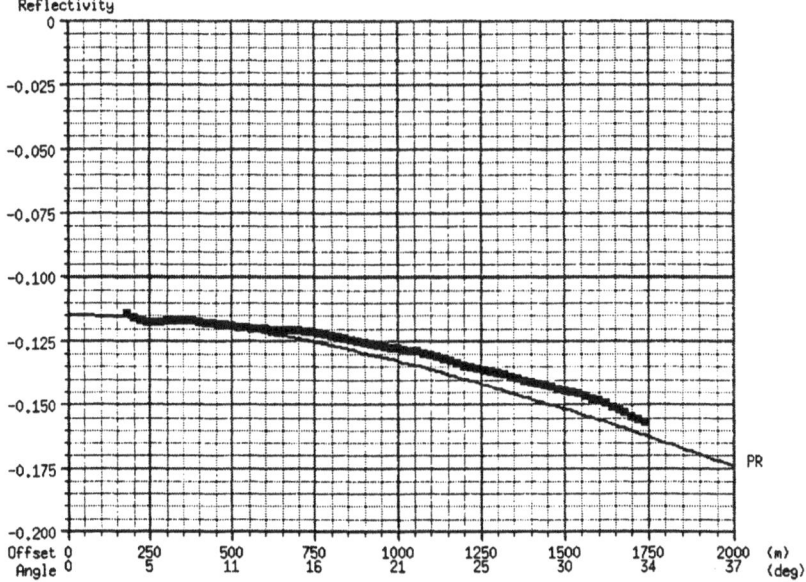

Fig. 53. *AVO response for the inelastic ($Q_{int} = 200$), thinly-layered Hordaland model after compensation with the pseudo-effective Q_{eff} ($Q_{eff} = 180$).(This Figure has been taken, with kind permission, from Widmaier, 1996, p. 79.)*

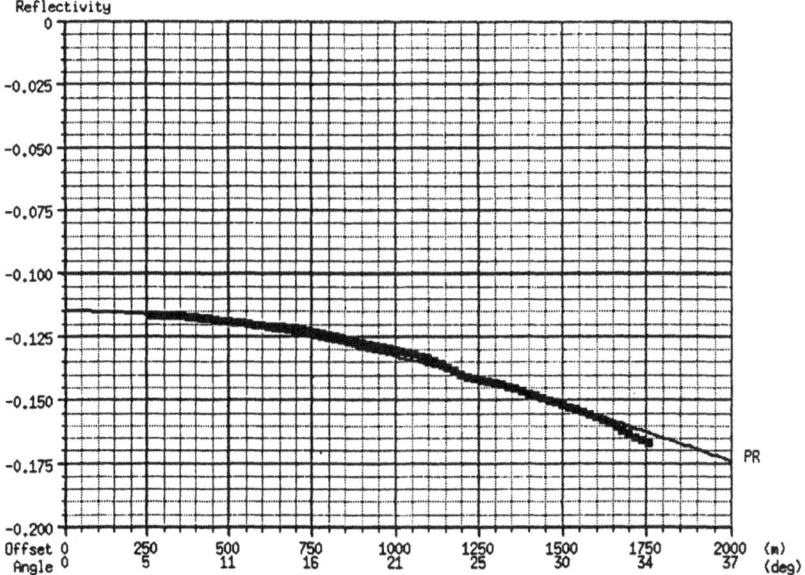

Fig. 54. *AVO response after Q-compensation ($Q_{int} = 200$) and thin-layering correction. (This Figure has been taken, with kind permission, from Widmaier, 1996, p. 79.)*

9 Stratigraphic Filtering in Poroelastic Media

A typical reservoir rock consists of *multi layered porous fluid saturated permeable sediments*. In such media the modeling of wave propagation can be done on the basis of the theory of poroelasticity (see the classical papers of M. Biot 1956a,b; 1962a,b). In this chapter we consider features of stratigraphic filtering in poroelastic vertically heterogeneous structures. The stratigraphic filtering includes now not only pure elastic scattering effects, but also inelastic effects due to the interaction of a viscous fluid with the porous matrix. Additionally, even for a vertical incidence of the P-wave (we restrict our consideration to this case) the wavefield is a multi-mode one.

In homogeneous poroelastic media the *dissipation* is caused by the relative movement between the solid matrix and the fluid. This mechanism is called the *global flow* and it was introduced by Biot (Biot, 1956, 1962). Biot's theory predicts the existence of *two compressional waves* in porous saturated rock, a normal P_1-mode and the highly attenuated P_2- mode, the so-called *slow wave*. In a heterogeneous medium, seismic waves propagating through a stack of layers with variable compliancies may additionally cause *inter-layer flow* of pore fluid (also termed *1-D local flow*) across interfaces from more compliant into stiffer layers and vice versa (White, 1983; Norris, 1993; Gurevich and Lopatnikov, 1995, Gelinsky and Shapiro, 1997a). The fluid pressure tends to equilibrate between adjacent layers due to viscous fluid motion across the layer boundaries. This can be explained as excitation of diffusive Biot-slow waves at the interfaces by scattering of the normal P-wave.

Thus, from one hand, in poroelastic fluid-saturated thinly-layered sediments the impedance fluctuations lead to elastic scattering. From another hand, variations of the layer compressibilities cause the inter-layer flow (a 1-D macroscopic local flow of the pore fluid). This results in significant attenuation and dispersion of the seismic wavefield even in the surface seismic frequency range, 10 - 100 Hz. The various attenuation mechanisms are approximately additive, dominated by the inter-layer flow at very low frequencies. Elastic scattering is important over a broad frequency range from seismic to sonic frequencies. Biot's global flow (the relative displacement of the solid frame and the fluid which is coherent on the

P-wave length scale) contributes mainly in the range of ultrasonic frequencies. From the seismic frequency range up to ultrasonic frequencies, attenuation due to heterogeneity is strongly enhanced compared to homogeneous Biot models.

Moreover, in homogeneous poroelastic systems with physical parameters realistic for usual hydrocarbon reservoirs the permeability tensor practically does not influence propagating seismic waves in the low frequency range (0-1000Hz). However, as we will see from the features of the poroelastic stratigraphic filtering, in heterogeneous systems the transport properties of rocks significantly affect seismic wavefields. Seismic signatures of the transport properties are essentially different in periodic structures and in media with disorder.

9.1   Model and Dynamic Equations

The starting point to describe poroelastic wave propagation is the system of Biot's equations (Biot, 1962), which we give here in the most general case of an anisotropic heterogeneous poroelastic medium:

$$
\sum_j \frac{\partial}{\partial x_j} \tau_{ij} = \frac{\partial^2}{\partial t^2}(\varrho u_i + \varrho_f w_i)
$$
$$
-\frac{\partial}{\partial x_i} p_f = \frac{\partial^2}{\partial t^2}\left(\varrho_f u_i + \sum_j m_{ij} w_j\right) + \eta \sum_j r_{ij} \frac{\partial}{\partial t} w_j,
$$

(9.1)

where τ_{ij} are total stress components in the reservoir rock, p_f is a fluid pressure, u_i are components of the solid matrix displacement, U_i are components of the average fluid displacement, φ is a porosity, η is a viscosity of the fluid, ϱ_f is a density of the fluid, ϱ is a density of the reservoir rock, m_{ij} are components of a symmetric tensor proportional to ϱ_f describing the micro geometry of the rock, and r_{ij} are components of the symmetric tensor of the flow resistivity, whereas the inverse of this tensor is the tensor of permeability. Finally, in the system above the relative displacement of the fluid has been introduced:

$$
w_i = \varphi(U_i - u_i).
$$

(9.2)

This system of equations must be completed by the stress-strain relations. For example, in the case of transverse isotropic rocks with the vertical axis (z) of symmetry we have:

$$
s = \begin{pmatrix}
2B_1 + B_2 & B_2 & B_3 & 0 & 0 & 0 & B_6 \\
B_2 & 2B_1 + B_2 & B_3 & 0 & 0 & 0 & B_6 \\
B_3 & B_3 & B_4 & 0 & 0 & 0 & B_7 \\
0 & 0 & 0 & B_5 & 0 & 0 & 0 \\
0 & 0 & 0 & 0 & B_5 & 0 & 0 \\
0 & 0 & 0 & 0 & 0 & B_1 & 0 \\
B_6 & B_6 & B_7 & 0 & 0 & 0 & B_8
\end{pmatrix} e,
\tag{9.3}
$$

where

$$
\begin{aligned}
&s = \left(\tau_{xx}, \tau_{yy}, \tau_{zz}, \tau_{yz}, \tau_{zx}, \tau_{xy}, p_f\right)^T, \\
&e = \left(e_{xx}, e_{yy}, e_{zz}, e_{yz}, e_{zx}, e_{xy}, \xi\right)^T, \\
&e_{ii} = \frac{\partial}{\partial x_i} u_i, \\
&e_{ij} = \frac{\partial}{\partial x_j} u_i + \frac{\partial}{\partial x_i} u_j, \quad (i \neq j), \\
&\xi = -\sum_i \frac{\partial}{\partial x_i} w_i.
\end{aligned}
\tag{9.4}
$$

The eight elastic coefficients B_i characterize the transverse isotropic reservoir rock. As other rock properties (like the permeability, viscosity, densities et al.) they should be experimentally determined (see e.g. Arts and Rasolofasaon, 1992). These elastic coefficients are related to elastic coefficients of the dry rock, of the material of the solid constituting the dry rock and compressibility of the fluid by the generalized Gassman formulas (Brown and Korringa, 1975).

Further, we consider the usual P-wave vertically propagating in a randomly-layered saturated (or partially saturated) structure. For the simplicity we assume that the statistics of heterogeneities is stationary and the standard deviations of physical parameters are small. We also will neglect the effects of density fluctuations, which are usually smaller than fluctuations of elastic properties.

Finally, we restrict our consideration to the case of isotropic layers.

As in the previous chapters we will use the following notations. We denote as ε_X average-normalized fluctuations of a corresponding poroelastic quantity X. By definition $\langle \varepsilon_X \rangle = 0$ and $\langle \varepsilon_X \varepsilon_Y \rangle = \sigma_{XY}^2$ is the covariance of the quantities X and Y. Clearly, σ_{XX}^2 and σ_{XX} stand for the variance and the standard deviation of the quantity X, respectively. Finally, the functions

$$\Phi_{XY}(\zeta) = \langle \varepsilon_X(z)\varepsilon_Y(z+\zeta) + \varepsilon_Y(z)\varepsilon_X(z+\zeta) \rangle / 2\sigma_{XY}^2$$

are called the correlation functions. For simplicity we assume that all correlation functions are equal to a single function $\Phi(\zeta)$ (this is a physically justified assumption).

For vertically propagating plane waves, equations (9.1) are transformed into the first-order differential matrix equations:

$$\frac{d\zeta}{dz} - \mathbf{P}\zeta = 0. \tag{9.5}$$

The wavefield parameters are contained in the vector $\zeta = (u_z, \tau_{zz}, w_z, p_f)^T$, and the matrix $\mathbf{P}$ is given by:

$$\mathbf{P} = \begin{pmatrix} 0 & \frac{1}{P_d} & 0 & \frac{\alpha}{P_d} \\ -\varrho\omega^2 & 0 & -\varrho_f\omega^2 & 0 \\ 0 & -\frac{\alpha}{P_d} & 0 & -\frac{1}{M} - \frac{\alpha^2}{P_d} \\ \varrho_f\omega^2 & 0 & i\omega\tilde{q} & 0 \end{pmatrix}, \tag{9.6}$$

where the following definitions have been introduced:

$$P_d = K_d + \frac{4}{3}\mu_d,$$
$$H = P_d + \alpha^2 M,$$
$$\alpha = 1 - \frac{K_d}{K_g}, \tag{9.7}$$
$$M = \left[\frac{\varphi}{K_f} + \frac{\alpha - \varphi}{K_g} \right]^{-1}.$$

Additionally, K_d and μ_d are the bulk and shear modulus, (throughout this chapter, the subscripts d, g, f denote properties of the dry frame, of the grain material, and of the fluid, respectively). Moreover, P_d is the dry P-wave modulus, and H is the saturated P-wave modulus which is called sometimes the *Gassman modulus*. Finally, the quantity $\tilde{q} = \eta/k$ is the Darcy-coefficient, and k is the permeability, which is assumed here to be a scalar (as we restrict our consideration to the case of isotropic layers).

9.2 Time-Harmonic Transmissivity

The further derivation is absolutely analogous to that, given in Chapter 5 (for more details see Gelinsky and Shapiro, 1997). These derivations provide us with the following result for the time-harmonic transmissivity of the usual compressional wave propagating through a stack of thin layers, which has a total thickness L and lies between two homogeneous halfspaces on its top and bottom:

$$T = \exp\left[i\left(\Psi L - \omega t\right) - \gamma L\right], \tag{9.8}$$

where $V_P = \omega/\Psi$ is the P- wave phase velocity, and γ is the attenuation coefficient. Both, Ψ and γ, are functions of the auto- and crosscorrelations of the medium fluctuations:

$$\Psi = \kappa_1^R + A - \int_0^\infty d\zeta\, \Phi(\zeta) \left[\sqrt{2}B \exp\left(-\zeta\kappa_-^I\right)\cos(\zeta\kappa_-^R - \frac{\pi}{4})\right.$$

$$\left. +\sqrt{2}B \exp\left(-\zeta\kappa_+^I\right)\cos(\zeta\kappa_+^R - \frac{\pi}{4}) + C\exp\left(-2\zeta\kappa_1^I\right)\sin(2\zeta\kappa_1^R)\right], \tag{9.9}$$

$$\gamma = \kappa_1^I + \int_0^\infty d\zeta\, \Phi(\zeta) \left[-\sqrt{2}B \exp\left(-\zeta\kappa_-^I\right)\cos(\zeta\kappa_-^R + \frac{\pi}{4})\right.$$

$$\left. -\sqrt{2}B \exp\left(-\zeta\kappa_+^I\right)\cos(\zeta\kappa_+^R + \frac{\pi}{4}) + C\exp\left(-2\zeta\kappa_1^I\right)\cos(2\zeta\kappa_1^R)\right]. \tag{9.10}$$

In results (9.9) and (9.10) we introduced the following notations: κ_1 and κ_2 are wavenumbers of the first (usual) and second (slow) compressional waves respectively; $\kappa_- = \kappa_2 - \kappa_1$, $\kappa_+ = \kappa_2 + \kappa_1$. The superscripts R and I define the real- and imaginary parts of the wavenumbers, e.g., $\kappa_1 = \kappa_1^R + i\kappa_1^I$. The three new quantities, A, B, and C are combinations of the variances of the medium fluctuations. We assume that there is only one correlation function $\Phi(\zeta)$ with correlation length l for all fluctuations. Our results are valid under the assumption of small fluctuations, neglecting terms of $O(\varepsilon^3)$ and higher, but they are valid for any relationship between the wavelength and the correlation length of the medium fluctuations and thus define *dynamic-equivalent-medium parameters*. Eqs. (9.9) and (9.10) are independent of any kind of low-frequency assumption even with respect to Biot's critical frequency.

Let us now explicitly write down the results for an exponential medium. The phase velocity can be found from:

$$\frac{\omega}{V_P} = \kappa_1^R + A - \frac{Bl(1 + l(\kappa_-^R + \kappa_-^I))}{1 + 2l\kappa_-^I + l^2(\kappa_-^{R^2} + \kappa_-^{I^2})} - \frac{Bl(1 + l(\kappa_+^R + \kappa_+^I))}{1 + 2l\kappa_+^I + l^2(\kappa_+^{R^2} + \kappa_+^{I^2})}$$
$$- \frac{2Cl^2\kappa_1^R}{1 + 4l\kappa_1^I + 4l^2(\kappa_1^{R^2} + \kappa_1^{I^2})}.$$
(9.11)

The corresponding attenuation coefficient (integration of Eq. (9.10)) reads:

$$\gamma = \kappa_1^I + \frac{Bl(1 - l(\kappa_-^R - \kappa_-^I))}{1 + 2l\kappa_-^I + l^2(\kappa_-^{R^2} + \kappa_-^{I^2})} + \frac{Bl(1 - l(\kappa_+^R - \kappa_+^I))}{1 + 2l\kappa_+^I + l^2(\kappa_+^{R^2} + \kappa_+^{I^2})}$$
$$+ \frac{Cl(1 + 2l\kappa_1^I)}{1 + 4l\kappa_1^I + 4l^2(\kappa_1^{R^2} + \kappa_1^{I^2})}.$$
(9.12)

Whereas the wavenumbers of Biot's fast and slow wave are well known for all frequencies (see below), the new constants A, B, and C are rather complicated functions of frequency and of the numerous auto- and crosscorrelation functions.

The last equation shows that there exist three physically different contributions to the process of the attenuation of the transmitted wavefield: (i) global flow, which corresponds to the imaginary part of the wavenumber κ_1; (2) elastic scattering ($P_1 \rightarrow P_1$), which corresponds to the last term in equation (9.12); (3) interlayer flow, which is described by the two similar terms with the factor B. All three effects are additive in the small-fluctuation approximation.

9.3   Below the Critical Frequency

In the seismic frequency range only two effects from the above mentioned three ones are of importance for the attenuation and velocity dispersion. The first one is the inter-layer (or local) flow, which is a fluid motion at interfaces in the structures. This effect is due to generation of the dissipative slow waves by scattering of seismic waves on heterogeneities. It can also be understood as the process of diffusion (i.e., pressure relaxation) on the scale of heterogeneities. The second important effect is the usual elastic scattering of seismic waves. The Biot global flow does not provide any significant contribution to the inelastic (or apparent inelastic) behavior of the system in the low frequency range.

Of course, in realistic porous rocks there are also other relaxation mechanisms, the description of which is beyond the Biot theory. Such an effect is the pressure-relaxation fluid motion on the single-grain (or pore) scales, which is called the squirt flow (see e.g., Dvorkin et al, 1994; Murphy et al, 1986 and references there). The magnitude and the frequency range of such mechanisms have not yet sufficiently been studied till now. They seem to be important for the frequency range of many kHz.

Thus, for seismic waves in heterogeneous poroelastic media an additional simplification is possible. We can consider an approximate solution, valid in the frequency range below Biot's critical frequency

$$\omega_c = \frac{\eta\varphi}{k\varrho_f}.$$
(9.13)

Assuming that $\varsigma = \omega/\omega_c \ll 1$, the small parameter ς can be used for a series expansions of the constants A, B, and C. Since the attenuation and velocity dispersion due to inter-layer flow and scattering are observable in most media for frequencies $\omega \ll \omega_c$, it is useful to neglect terms of $O(\varsigma^3)$ and higher in the various series expansions. Since for $\varsigma \ll 1$ it always holds that

$$\kappa_2^R \gg \kappa_1^R \gg \kappa_1^I = \frac{\omega^2}{2\omega_c}\sqrt{\frac{\varrho}{H}}\frac{\varphi\varrho}{\varrho_f}\left(\frac{\varrho_f}{\varrho} - \frac{\alpha M}{H}\right)^2,$$
(9.14)

we obtain that $\kappa_+ \approx \kappa_- \approx \kappa_2$ and

$$\kappa_2 \approx (1+i)\sqrt{\frac{\omega\eta}{2kN}} = (1+i)\sqrt{\frac{\omega\omega_c\varrho_f}{2\varphi N}},$$

$$\kappa_1 \approx \kappa_1^R \approx \omega\sqrt{\frac{\varrho}{H}}, \quad N = \frac{MP_d}{H}. \tag{9.15}$$

In the following we will understand under κ_2 the factor $\sqrt{\omega\eta/2kN}$ only. Instead of the attenuation coefficient we give already the quality factor $Q^{-1} = 2\gamma/\kappa_1$. Thus, equations (9.11) and (9.12) give

$$\frac{\omega}{V_P} = \kappa_1 + \kappa_1 A' - 2\frac{\kappa_1\kappa_2 B'l(1+2l\kappa_2)}{1+2l\kappa_2+2l^2\kappa_2^2} - \frac{2C'l^2\kappa_1^2}{1+4l^2\kappa_1^2},$$

$$Q^{-1} = 4\frac{B'\kappa_2l(1-2l\kappa_2)}{1+2l\kappa_2+2l^2\kappa_2^2} + 2\frac{\kappa_1 C'l}{1+4l^2\kappa_1^2}, \tag{9.16}$$

where A', B' and C' are new frequency-independent combinations of variances of the medium parameters. These equations show the following characteristic frequencies for the inter-layer flow and the poroelastic scattering:

$$\omega_{flow} = \frac{2Nk}{l^2\eta}, \quad \omega_{scat} = \frac{1}{2l}\sqrt{\frac{H}{\varrho}}. \tag{9.17}$$

Usually, $\omega_{flow}, \omega_{scat} \ll \omega_c$.

Keeping in mind the frequency dependencies of $\kappa_1 \propto \omega$, and $\kappa_2 \propto \sqrt{\omega}$, it is easy to see that the quality factor Q has two relaxation peaks at the characteristic frequencies given above. Respectively, the phase velocity has three limiting values that may be called a quasi-static velocity V_{qs} for $\omega \to 0$, an intermediate no-flow velocity V_{nf} for $\omega \gg \omega_{flow}$ but wavelengths smaller than the correlation length l, and finally a ray-theoretical limit V_{ray} above that limit (but necessarily below Biot's ω_c).

For illustration we show here the velocities and quality factors consisting of a heterogeneous stack of layers with average properties of water saturated Berea sandstone (Norris, 1993). We assume that the medium exhibits fluctuations of

P_d, of α, and of its fluid saturation contained in M:

$$P_d = P_0(1 + \varepsilon_p(z)),$$
$$\alpha = \alpha_0(1 + \varepsilon_\alpha(z)), \qquad\qquad (9.18)$$
$$M = M_0(1 + \varepsilon_M(z)).$$

For these fluctuations the constants A', B', and C' are:

$$A' = \frac{1}{2}\frac{P_0}{H_0}\left(\sigma_{PP}^2 + \frac{\alpha_0^2 M_0}{P_0}(\sigma_{MM}^2 + 2\sigma_{P\alpha}^2) + \frac{\alpha_0^4 M_0^2}{P_0^2}\sigma_{\alpha\alpha}^2\right),$$

$$B' = \frac{1}{4}\frac{P_0\alpha_0^2 M_0}{H_0^2}[\sigma_{PP}^2 - 2\sigma_{PM}^2 + \sigma_{MM}^2 -$$

$$- 2\frac{P_0 - \alpha_0^2 M_0}{P_0}(\sigma_{P\alpha}^2 - \sigma_{M\alpha}^2) + \frac{(P_0 - \alpha_0^2 M_0)^2}{P_0^2}\sigma_{\alpha\alpha}^2], \qquad\qquad (9.19)$$

$$C' = \frac{1}{4}\frac{P_0^2}{H_0^2}[\sigma_{PP}^2 + 2\frac{\alpha_0^2 M_0}{P_0}(\sigma_{PM}^2 + 2\sigma_{P\alpha}^2) +$$

$$+ \frac{\alpha_0^4 M_0^2}{P_0^2}(\sigma_{MM}^2 + 4\sigma_{\alpha\alpha}^2 + 4\sigma_{M\alpha}^2)].$$

In our model P- velocity fluctuations of almost 15% and the existence of zones with partial gas saturation (2% gas) yield the following variances (constant density): $\sigma_{PP}^2 = 0.080$, $\sigma_{MM}^2 = 0.065$, $\sigma_{\alpha\alpha}^2 = 0.004$, $\sigma_{PM}^2 = -0.0018$, $\sigma_{M\alpha}^2 = 0.0002$, $\sigma_{P\alpha}^2 = -0.018$. The corresponding average values are: $P_0 = 2.95 \cdot 10^{10}$Pa, $M_0 = 7.33 \cdot 10^9$Pa, and $\alpha_0 = 0.79$. The resulting phase velocity is plotted in Figure 55 and the attenuation Q^{-1} in Figure 56, both as functions of frequency, normalized by Biot's critical frequency $\omega_c = 2\pi \cdot 1.6 \cdot 10^5 \text{s}^{-1}$. The correlation lengths considered in Figures 55 and 56 range from 3 m down to 3 cm (curves from left to right: $l = 3$ m, 1 m, 0.3 m, 0.1 m, and 0.03 m). For the thickest layers ($l = 3.0$ m, 1.0 m) the inter-layer flow cannot equilibrate the pressure at seismic frequencies - the maximum of inter-layer flow attenuation is below 1 Hz. For the thinnest layers ($l = 0.1$ m, 0.03 m) there is a continuous change from the quasi-static to the ray-theoretical velocity limit. This high-frequency limit is slightly affected by global-flow dispersion. For the larger correlation lengths, the peaks of scattering and inter-layer flow can be distinguished. Attenuation due to Biot-global-flow becomes relevant above $\omega/\omega_c \approx 10^{-2}$.

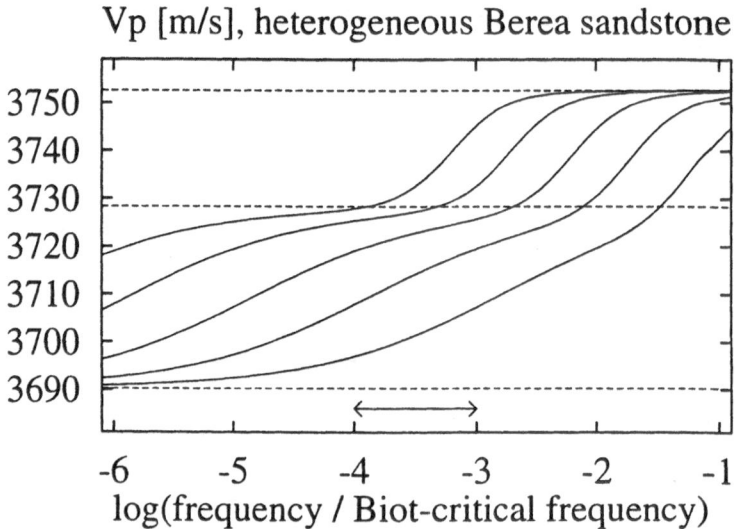

Fig. 55. Phase velocity as function of normalized frequency ($f_c = 1.6 \cdot 10^5$ Hz) for various correlation lengths. V_P was calculated for a heterogeneous stack of layers with average properties of Berea sandstone. The dashed lines show the limiting velocities, from top to bottom: V_{ray}, V_{nf}, and V_{qs}. The solid lines were calculated for correlation lengths $l = 3.0$ m, 1.0 m, 0.3 m, 0.1 m, 0.03 m (from upper left to lower right). The little arrows mark frequency range 15-150 Hz. (Reprinted from Gelinsky and Shapiro, 1997, with the kind permission of Blackwell Science.)

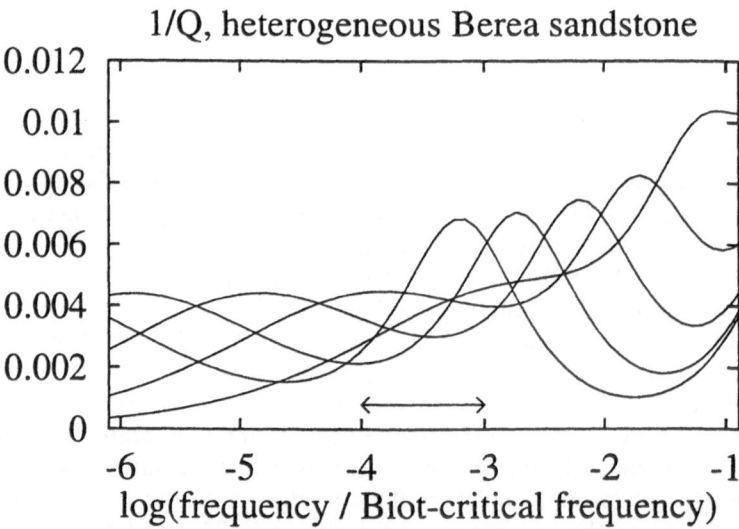

Fig. 56. Reciprocal quality factor Q^{-1} for the same model as in Figure 55. From the left to right peak the correlation length is: $l = 3.0$ m, 1.0 m, 0.3 m, 0.1 m, 0.03 m. (Reprinted from Gelinsky and Shapiro, 1997, with the kind permission of Blackwell Science.)

9.4   Attenuation and Permeability

In homogeneous poroelastic systems the permeability tensor practically does not influence the propagating seismic waves in the low frequency range (0-1000Hz; see, e.g., Gelinsky and Shapiro, 1996 or Schmitt, 1989). This situation changes in heterogeneous systems, like, e.g., layered or fractured sediments. Due to the heterogeneities of poroelastic structures the attenuation of P-waves is influenced by the permeability in an enhanced way. Following Shapiro and Müller (1997), we show here that such *a 'seismic permeability' can differ very strongly from the hydraulic permeability.*

We again return to the general equation (9.10). Neglecting the global-flow effects we obtain the following formula for the reciprocal quality factor Q^{-1}:

$$Q^{-1} = \frac{\alpha_0^2 N_0 \Delta_1}{H_0} F_{flow}(x_{flow}) + \frac{N_0^2 \Delta_2}{M_0^2} F_{scat}(x_{scat}). \tag{9.20}$$

Quantities α_0, M_0, N_0, and H_0 characterize the averaged properties of the medium. In contrast to them quantities $\Delta_{1,2}$ are measures of heterogeneity of the medium. They are two different linear combinations of the normalized variances and covariances of the poroelastic parameters. For example, in the case of fluctuating P_d, α, and M

$$\Delta_1 = \left\langle \left(\varepsilon_P - \varepsilon_M - 2\frac{P_0 - \alpha_0^2 M_0}{P_0}\varepsilon_\alpha \right)^2 \right\rangle,$$

$$\Delta_2 = \sigma_{HH}^2. \tag{9.21}$$

The factors before the functions F do not contain any dynamic information (i.e., frequency ω). Accordingly, they do not contain any information on the permeability. The dynamic dependence of the attenuation is contained in the quantities $F_{flow}(x)$ and $F_{scat}(x)$. These are positive functions of the following form:

$$F_{flow}(x_{flow}) = \sqrt{2}x_{flow} \int_0^\infty d\zeta \Phi(\zeta) \exp(-\zeta x_{flow}) \cos(\zeta x_{flow} + \pi/4)$$

$$F_{scat}(x_{scat}) = \frac{1}{2}x_{scat} \int_0^\infty d\zeta \Phi(\zeta) \cos(2\zeta x_{scat}). \tag{9.22}$$

Here $\zeta = \Delta z / l$ and Δz is the correlation lag. For simplicity we assume that all correlation functions are equal to a single function $\Phi(\zeta)$. The arguments of the functions F_{flow} and F_{scat} depend on frequency:

$$x_{flow} = \sqrt{\omega a^2 r / 2N_0}; \quad x_{scat} = \omega a / c_0, \tag{9.23}$$

where $c_0 = \sqrt{H_0 / \varrho_0}$ is the low-frequency limit of the P-wave velocity, ϱ_0 is the averaged density of the saturated rock and $r = \eta / k$ (for the moment we assume this quantity to be a constant).

The second term in equation (9.20) describes the usual elastic scattering. This term does not depend on the permeability and, therefore, in the following analysis we mainly consider the contribution of the inter-layer flow (given by the first term).

The values of Δ_1 and Δ_2 in equation (9.20) are usually of the order of $0.01 - 0.1$. Under the assumptions $K_f \ll \varphi K_g$, and accepting that fluctuations of the K_g are much smaller than the fluctuations of other parameters we obtain the following rough approximation

$$\Delta_1 \approx \langle (\varepsilon_P - \varepsilon_{Kf} + \varepsilon_\varphi)^2 \rangle. \tag{9.24}$$

This relation shows the role of cross-correlations between the fluctuations. For instance, if the fluctuations of the skeleton elastic modulus and the fluctuations of the bulk modulus of the pore fluid are anti correlated, then a maximum inter-layer flow contribution in the attenuation of the P-wave will be observed. Such a situation is possible in partially saturated reservoirs. In realistic situations the low-frequency range attenuation due to the inter-layer flow can be of the order of $n * 100$ ($1 < n < 9$) in terms of the Q-factor. The contribution of the elastic scattering is also of the same order. In extreme cases (of partial gas saturation) the resulting Q can reach the order of $n * 10$.

The frequency and permeability dependencies of the attenuation are defined by functions $F_{flow}(\bar{x})$ and $F_{scat}(x)$. Therefore, they are controlled by the correlation properties of the medium heterogeneities (i.e., by the disorder of rock structures). Let us first consider a periodic structure (e.g., no disorder). Taking into account symmetry properties of the correlation function we find in the low frequency limit (see Appendix 9.5 for a detailed calculation).

$$F_{flow} \propto \omega; \quad F_{flow} \propto k^{-1}. \tag{9.25}$$

Therefore Q^{-1} is proportional to frequency and the attenuation coefficient is proportional to ω^2 and inversely proportional to the permeability. The high-frequency limit of the function F_{flow} depends on individual properties of the function Φ.

Let us now consider the general case of randomly heterogeneous media (i.e., media with disorder). In this case the correlation between any two locations in the medium decreases with an increasing distance between them. Therefore, the function $\Phi(\zeta)$ is a quickly decreasing function for $\zeta > 1$. Taking this into account and considering the limits $x_{flow} \longrightarrow 0$ and $x_{scat} \longrightarrow 0$ we obtain

$$F_{flow}(x_{flow}) \approx x_{flow} \int_0^1 d\zeta \Phi(\zeta); \qquad F_{scat}(x_{scat}) \approx \frac{1}{2} x_{scat} \int_0^1 d\zeta \Phi(\zeta). \qquad (9.26)$$

Therefore, we observe that in any medium with disorder, in the low-frequency range the Q^{-1}-contribution of the inter-layer flow is proportional to x_{flow}, i.e., to $\omega^{1/2}$. The contribution of the elastic scattering is proportional to ω. Thus, the inter-layer-flow associated attenuation coefficient is proportional to $\omega^{3/2}$. This frequency dependence is analogous to the Rayleigh-scattering frequency dependence and it is universal for all poroelastic media with disorder. Moreover, the low-frequency-range attenuation coefficient is proportional to $k^{-1/2}$.

There is no universal behavior in the case of $x_{flow} \gg 1$, where the frequency dependence of the inter-layer-flow attenuation is controlled by the statistics of the heterogeneities. Let us consider a rather general case of fractal-like heterogeneities (including exponentially correlated ones) characterized by the von Karman correlation function

$$\Phi(\zeta) = 2^{1-\nu} \Gamma^{-1}(\nu) \zeta^\nu K_\nu(\zeta), \qquad (9.27)$$

where K_ν is the modified Bessel function of third kind (Macdonald function). We find that in the high frequency limit (see Appendix 9.6):

$$F_{flow} \propto x_{flow}^{-2\nu}, \qquad F_{scat} \propto x_{scat}^{-2\nu}. \qquad (9.28)$$

In spite of their similarity these relations result in different frequency dependencies of the fluid-flow and scattering contributions to the attenuation:

$$Q_{flow}^{-1} \propto \omega^{-\nu}, \qquad Q_{scat}^{-1} \propto \omega^{-2\nu}. \qquad (9.29)$$

In addition, the permeability dependence reads:

$$Q_{flow}^{-1} \propto k^{\nu}. \tag{9.30}$$

As a particular example we consider the case $\nu = 1/2$. Then the correlation function assumes the exponential form $\Phi(\zeta) = \exp(-\zeta)$ and the functions F read:

$$F_{flow}(x_{flow}) = \frac{x_{flow}}{1 + 2x_{flow} + 2x_{flow}^2}, \qquad F_{scat}(x_{scat}) = \frac{x_{scat}/2}{1 + 4x_{scat}^2}. \tag{9.31}$$

This example shows that functions $F_{flow}(x)$ and $F_{scat}(x)$ are positive functions with magnitude of the order $O(1)$. These functions reach their maxima at $x = O(1)$. Therefore, the permeability controls the location of the maximum of the inter-layer-flow part in the frequency range. This maximum is reached at frequencies of the order $O(2Nk/l^2\eta)$. In realistic situations, if l is in the range of $10^{-2} - 10m$ the maximum of the Q^{-1} factor of the inter-layer-flow attenuation can be reached at any frequency lower than 10^3 Hz. Therefore this attenuation will be significant at least in a part of the seismic frequency range.

Because the typical order of l is about of 1m, in the seismic frequency range the high-frequency branch of the inter-layer flow is more likely to be observed. Thus, it should be expected that a larger attenuation corresponds to a larger permeability in the case of equivalent other conditions. Moreover, the wave propagation along a crack system with a larger permeability should be accompanied by a larger attenuation than the propagation along crack systems with smaller permeability. Thus, an anisotropy of attenuation due to the permeability anisotropy can be observed.

In reality the permeability is a strongly fluctuating quantity. Therefore, the permeability must enter into equations (9.22) in a somehow averaged form. It is well known that in the static limit the permeability of 1-D heterogeneous media in the direction normal to the layering is given by the averaging $\langle 1/k \rangle^{-1}$. This permeability controls a large-scale fluid flow through the layered system (see e.g., Gelinsky and Shapiro, 1997). It is important for computation of hydraulic properties of reservoirs. We call it *the hydraulic permeability*.

However, numerical simulations show that this estimation of k does not control the seismic attenuation. Indeed, in Figure 57 the results of numerical simulations of the attenuation of the transmitted normal P-wave are given. The maximum in the lower frequency range is due to the inter-layer flow. The right-hand maximum is due to the scattering. The medium is characterized by the following average parameters: $c_0 = 3600m/s, \varrho = 2.4g/cm^3, \varphi = 0.15$. The fluctuations are exponentially correlated with $l = 1m$ and $\sigma_{Kf} = 0.2, \sigma_P = 0.3, \sigma_\varphi = 0.05$. The permeability was normally distributed with the standard deviation of 120 per cent. Thus, the layer permeability range from $1\mu D$ up to $1D$. For computation of the theoretical curve the harmonic averaged value of the permeability $k = 7\mu D$ (i.e., the hydraulic permeability) was used. It is clear that the hydraulic permeability cannot be used to describe the seismic attenuation correctly.

This is explained by the following: the inter-layer flow is a local flow effect, where the fluid motion takes place not through the total system but rather around heterogeneities in regions of the scale of the slow-wave wavelength. Such regions provide contributions to the dissipation of the elastic energy. These contributions are controlled by local values of the permeability. The propagating P-wave averages then all contributions. Therefore, in the case of a weak correlation between the permeability fluctuations and fluctuations of other poroelastic parameters the following rule should be applied:

$$Q_{ef}^{-1} = \int_0^\infty Q^{-1}(k)f(k)dk, \tag{9.32}$$

where Q_{ef}^{-1} is the resulting inverse quality factor, $Q^{-1}(k)$ is given by equation (9.20) and $f(k)$ is the probability density of k. Figure 57 clearly shows that this estimation of Q^{-1} is in good agreement with the numerical experiment. Therefore, the value of the permeability, which controls the seismic attenuation (we call this quantity *seismic permeability* and denote it as k_s) can be estimated from the following equation:

$$Q^{-1}(k_s) = Q_{ef}^{-1}, \tag{9.33}$$

For normally distributed k the following estimation can be satisfactory:

$$k_s = \langle k \rangle. \tag{9.34}$$

Such an averaging rule was suggested by Berryman (1986) for the case of the attenuation caused by the global flow. As we can see here the description of the attenuation in the case of local flow requires a more sophisticated rule of permeability averaging than in the case of the global flow. This rule should also take into account an information on cross-correlation between different parameters defining Q^{-1} (like, e.g., variances of the poroelastic modulus) and the

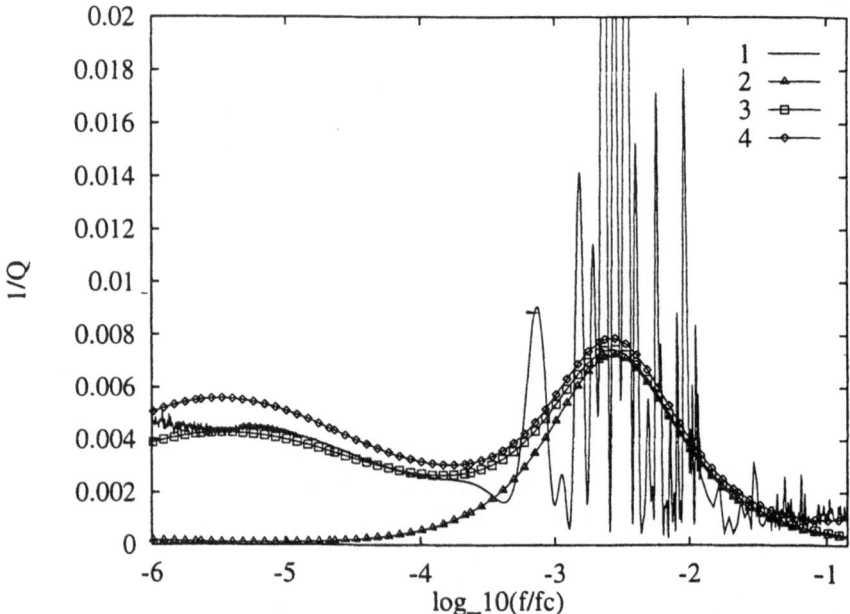

Fig. 57. Inverse quality factor of the P-wave versus frequency in a randomly layered porous water-saturated (with a fluctuating gas concentration) medium. Curve 1 shows results of numerical simulations (matrix-propagator method for Biot equations; see Schmidt and Tango, 1986). Curve 2 shows results of equation (9.20), where the hydraulic permeability was used. Curve 3 shows results of the averaging rule (9.32). Normal averaging yields an average permeability $k = 300mD$ (this corresponds to a Biot critical frequency of $f_c = 81kHz$), which after substitution into equation (9.20) provides curve 4.

permeability. Therefore, even in media with huge fluctuations of the permeability the results described here are applicable and the seismic signatures of the transport properties can be observed. It is, however, evident that the permeability controlling the seismic attenuation differs very strongly from the hydraulic permeability. This difference can reach several orders of magnitude.

9.5   Appendix: Periodic Media – Low-Frequency Limit

In periodic structures the correlation function $\Phi(\zeta)$ is periodic with the period 1. Thus, the first equation from (9.22) provides:

$$F_{flow}(x_{flow}) = \tag{9.35}$$

$$= \sqrt{2}x_{flow} \sum_{n=0}^{\infty} e^{-nx_{flow}} \int_0^1 d\zeta \Phi(\zeta) \exp(-\zeta x_{flow}) \cos((\zeta + n)x_{flow} + \pi/4)$$

$$= \sqrt{2}x_{flow} \int_0^1 d\zeta \Phi(\zeta) \exp(-\zeta x_{flow}) C_1(\zeta, x_{flow}), \tag{9.36}$$

where

$$C_1(\zeta, x_{flow}) = \\ = \frac{\cos(\zeta x_{flow} + \pi/4) - \exp(-x_{flow}) \cos(x_{flow} - \zeta x_{flow} - \pi/4)}{1 - 2\exp(-x_{flow}) \cos x_{flow} + \exp(-2x_{flow})}. \tag{9.37}$$

In order to proceed further with our analysis we must take into account the following property of a periodic correlation function:

$$\Phi(\zeta) = \Phi(-\zeta) = \Phi(1 - \zeta). \tag{9.38}$$

Using this property we obtain

$$F_{flow}(x_{flow}) = \sqrt{2}x_{flow} \int_0^{1/2} d\zeta \Phi(\zeta) C_1(\zeta, x_{flow}), \tag{9.39}$$

where

$$\begin{aligned} C_1(\zeta, x_{flow}) = &[\cos(\zeta x_{flow} + \pi/4) \exp(-x_{flow}\zeta) - \\ &- \cos(x_{flow} - \zeta x_{flow} - \pi/4) \exp(-x_{flow} - x_{flow}\zeta) + \\ &+ \cos(x_{flow} - \zeta x_{flow} + \pi/4) \exp(x_{flow}\zeta - x_{flow}) - \\ &- \cos(\zeta x_{flow} - \pi/4) \exp(x_{flow}\zeta - 2x_{flow})] / \\ &[1 - 2\exp(-x_{flow}) \cos x_{flow} + \exp(-2x_{flow})]. \end{aligned} \tag{9.40}$$

Considering now the limit $x_{flow} \longrightarrow 0$ we obtain

$$F_{flow}(x_{flow}) \propto x_{flow}^2. \tag{9.41}$$

9.6   Appendix: Random Media – High-Frequency Limit

Substituting the von Karman correlation function (9.27) into equations (9.22) we obtain

$$F_{flow}(x_{flow}) = 2^{\nu-1/2}\nu\Gamma(\nu+1/2)z\left[(1+2ix_{flow}^2)^{\frac{-2\nu-1}{4}}P_{-1/2+\nu}^{-1/2-\nu}(\bar{z})-\right.$$
$$\left.-i(1-2ix_{flow}^2)^{\frac{-2\nu-1}{4}}P_{-1/2+\nu}^{-1/2-\nu}(z)\right] \tag{9.42}$$

$$F_{scat}(x_{scat}) = \sqrt{\pi}\Gamma(\nu+1/2)\Gamma^{-1}(\nu)x_{scat}(1+x_{scat}^2)^{-1/2-\nu}, \tag{9.43}$$

where the function $P_{-1/2-\nu}^{-1/2-\nu}$ is the associated Legendre function of the first kind, $z = x_{flow} + ix_{flow}$, and $\bar{z}$ its complex conjugated. In order to derive the first result we used the tables of integrals (Prudnikov et al., 1988, p.349, eq. 2.16.6.3). In the high-frequency limit ($x_{flow,scat} \longrightarrow \infty$) we obtain equation (9.28)

10   Reflectivities of Multilayered Structures

In contrast to the previous chapters, where properties of transmissivities were considered, here we review some properties of wavefields reflected by stratified structures. Simple theoretical expressions for the reflectivity of multilayered structures are of importance because of several reasons. Reflectivity is the main source of information about the subsurface in seismic exploration. Such important seismic-processing procedures like imaging, inversion, predictive deconvolution, multiples attenuation, etc deal with the reflectivity. Moreover, a signal reflected from a randomly layered structure carries information on its statistics. If this information could be extracted from the reflectivity it could be used for two different needs. First, if we have a statistical macro model of the small-scale fluctuations in the overburden of a target reflector then a correction for the stratigraphic filtering effect, as indicated in the previous chapters, could be performed without invoking well-log data. Second, many geological structures of interest are stratified. A simple analytic approximation for the reflectivity of a thinly layered stack can then help to design inversion techniques (e.g., similar to the AVO techniques) providing important lithologic information, like the statistics of such a structure, or even, the distribution of its elastic properties with the depth.

10.1   Normal-Incidence Plane Wave

In this section we return to the simplest model: the Goupillaud model. The analysis of the transmissivity given in Chapter 3 will help us to find an approximation of the reflectivity taking the effects of multiple scattering into account. Note, that in this section as well as in Chapter 3, the variable z denotes the argument of the z-transforms of the reflectivity and transmissivity. In other parts of this book z is the depth.

The recursive equations (3.6) and (3.8) provide us with the following exact expression for the reflectivity of a layered stack with $n + 1$ interfaces:

$$R(z) = \frac{c_0 + \sum_{i=1}^{n} c_i A_i^R(z)}{A_n(z)}. \tag{10.1}$$

However, due to the polynomial A_n^R, this equation cannot be used to obtain any reliable approximation. In order to be able to apply the approximations of the polynomial A_n developed in the Chapter 3 we must turn to the Hubral-Treitel-Gutowski formula (3.18). Substituting approximation (3.29) into formula (3.18) we obtain:

$$R(z) = c_0 + \sum_{j=1}^{n} z^j c_j (1 - c_0^2)(1 - c_1^2)...(1 - c_{j-1}^2) \times$$

$$\times \exp\left[-c_0 c_j z^j - 2 \sum_{k=1}^{j-1} \sum_{i=0}^{j-k} c_i c_{i+k} z^k \right]. \tag{10.2}$$

Figure 58 shows the δ-pulse reflectivity computed exactly for the reflection-coefficient series given in Figure 5. In the same plot the approximate reflectivity obtained from equation (10.2) is shown. A very good agreement between the exact result and the approximation is evident.

In contrast to the exact equation (3.18) all terms of formula (10.2) have a rather simple and clear structure. This formula also includes effects of multiple scattering (what can be clearly seen from a comparison of Figures 5 and 58). Therefore, approximation (10.2) can serve as a base for studying various useful properties of the seismograms.

Approximation (10.2) still requires intensive calculations. However, if we look upon the reflection coefficient series as a realization of a random stationary process we will arrive at the following hybrid (statistic/deterministic) approximation of the reflectivity:

$$R(z) = c_0 + \sum_{j=1}^{n} z^j c_j \exp\left[(1 - 2j)S_1(z) - \varrho_c(0)j \right], \tag{10.3}$$

where the quantity $S_1(z)$ is given by formula (3.39). A simple analysis of the expression in the exponent shows that in the case of weakly correlated reflection coefficients this expression is roughly of the order of $2j\varepsilon^2$, where ε is the RMS

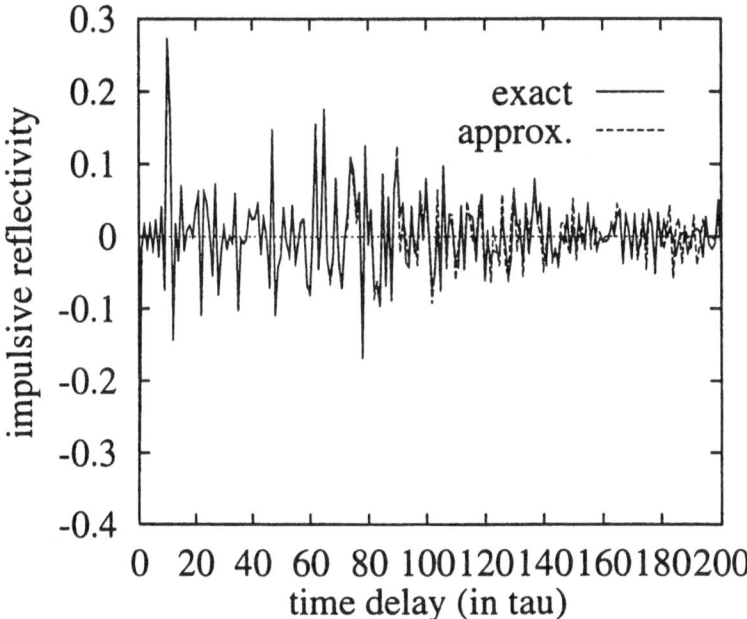

Fig. 58. The δ-pulse reflectivity for the reflection-coefficient series shown in Figure 5. The two curves shown are the exact reflectivity and the result of approximation (10.2), respectively. (Reprinted from Phys. of the Earth and Planet. Inter., v.104., p.156, 1997, Shapiro and Treitel, with kind permission from Elsevier Science - NL)

reflection coefficient. Therefore, first k terms of $R(z)$ practically coincide with $c_0, c_1z, ..., c_kz^k$, where $2k \ll \varepsilon^{-2}$. This can be immediately used for an inversion.

Based on equations (10.2) and (10.3) various inversion procedures can be developed. Such inversions should take into account effects of multiple scattering, which are always presented in real seismograms. In this respect, the inversion problem we speak about can be looked upon as a kind of dynamic predictive deconvolution (see Robinson, 1975).

In order to show that such an inversion can be useful for practical purposes we performed a very simple and approximate reconstruction of the first 100 reflection coefficients from the exact reflectivity given in Figure 58. We assumed again that the reflection-coefficient series is stationary and used equation (10.3) in the following heuristic way. We multiplied each lth sample of the reflected seismogram by the factor, which is reciprocal to the exponential factor from equation (10.3):

$$\exp\left[(2l-1)S_1(z) + \varrho_c(0)l\right]. \tag{10.4}$$

This factor depends on roughly estimated statistical properties of the reflection-coefficient series (Figure 5) only. Also for computing the quantity $S_1(z)$ the exponential autocorrelation of the acoustic impedance has been assumed. After this multiplication the inverse $z-$transform (i.e., the inverse Fourier transform) provides a series of reconstructed reflection coefficients.

Figure 59 shows (1) the first 100 reflection coefficients from Figure 5 and (2) the first 100 samples of the δ-pulse reflectivity computed exactly. Figure 60 shows (1) the same 100 first reflection coefficients from Figure 5, now however, in comparison with (2) the first 100 reflection coefficients reconstructed by the procedure described above. It is evident that the reconstructed reflection-coefficient series is closer to the true reflection coefficient series than the $\delta-$pulse reflectivity.

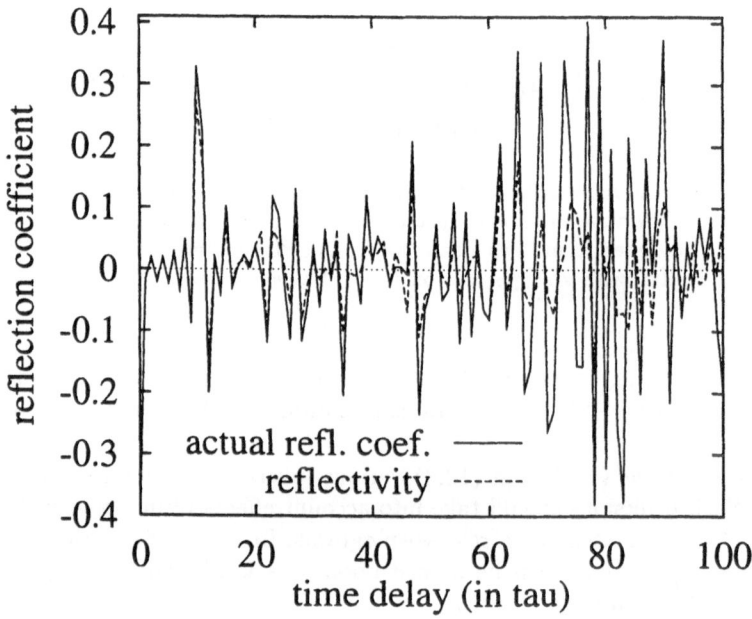

Fig. 59. The first 100 reflection coefficients shown in Figure 5 in comparison with the exact δ-pulse reflectivity. (Reprinted from Phys. of the Earth and Planet. Inter., v.104., p.157, 1997, Shapiro and Treitel, with kind permission from Elsevier Science - NL)

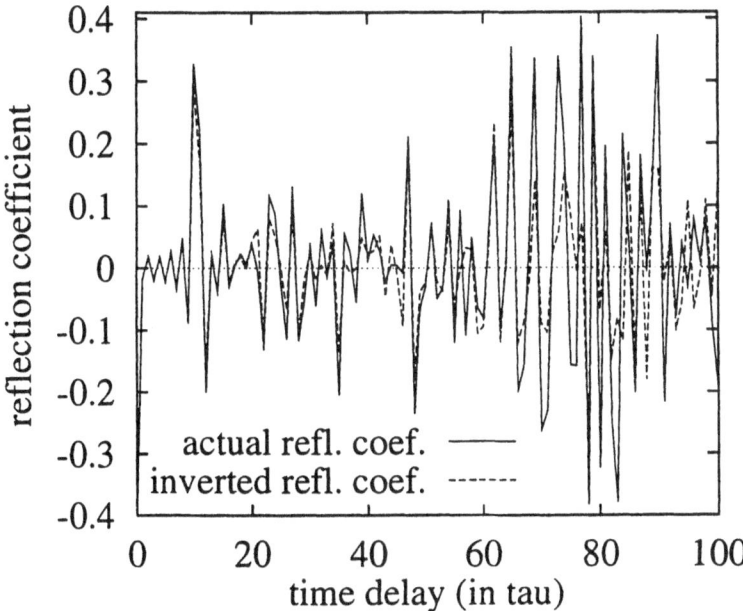

Fig. 60. The first 100 reflection coefficients shown in Figure 5 in comparison with the reflection coefficients heuristically reconstructed from the exact reflectivity. (Reprinted from Phys. of the Earth and Planet. Inter., v.104., p.157, 1997, Shapiro and Treitel, with kind permission from Elsevier Science - NL)

10.2   Oblique-Incidence Elastic Plane Waves

In order to derive the reflectivities of the P- and SV-waves, caused by a plane wave obliquely incident on a multilayered structure we must again return to the system of equations (5.23). We consider a plane wave incident in the uppermost half space (all notations are now the same as in Chapter 5). The corresponding reflectivities are given by eight equations additional to the eight equation of **system (5.27):**

$$\frac{\partial}{\partial \zeta} \Psi_{up} = r_{upp} C + r_{usp} D + A + t_{ups} e^{-i\kappa_a \zeta} E,$$

$$\frac{\partial}{\partial \zeta} t_{ups} = t_{ups} \left(i\kappa_b + F \right) + e^{i\kappa_a \zeta} \left(r_{uss} D + r_{ups} C + B + B\Psi_{up} \right),$$

$$\frac{\partial}{\partial \zeta} t_{usp} = t_{usp} \left(i\kappa_a + A \right) + e^{i\kappa_b \zeta} \left(E + E\Psi_{us} + r_{usp} H + r_{upp} G \right),$$

$$\frac{\partial}{\partial \zeta} \Psi_{us} = r_{ups} G + r_{uss} H + F + t_{usp} e^{-i\kappa_b \zeta} B,$$

$$\frac{\partial}{\partial \zeta} r_{dpp} = e^{2i\kappa_a \zeta + \Psi_{up} + \Psi_{dp}} C + e^{i\kappa_a \zeta + \Psi_{dp}} t_{ups} G + e^{i\kappa_a \zeta + \Psi_{up}} t_{dsp} D,$$

$$\frac{\partial}{\partial \zeta} r_{dps} = t_{dps} e^{i\kappa_a \zeta + \Psi_{up}} C + t_{ups} e^{i\kappa_b \zeta + \Psi_{ds}} H + e^{i(\kappa_b + \kappa_a)\zeta + \Psi_{ds} + \Psi_{up}} D,$$

$$\frac{\partial}{\partial \zeta} r_{dsp} = e^{i\kappa_a \zeta + \Psi_{dp}} t_{usp} C + e^{i(\kappa_a + \kappa_b)\zeta + \Psi_{dp} + \Psi_{us}} G + t_{dsp} e^{i\kappa_b \zeta + \Psi_{us}} H,$$

$$\frac{\partial}{\partial \zeta} r_{dss} = t_{dps} e^{i\kappa_b \zeta + \Psi_{us}} G + e^{i\kappa_b \zeta + \Psi_{ds}} t_{usp} D + e^{2i\kappa_b \zeta + \Psi_{ds} + \Psi_{us}} H,$$

$$(10.5)$$

where the terms smaller than second order of the medium fluctuation have been neglected. In the equations for the reflection coefficients, we have not expanded the exponential factors. This is helpful for an analysis of the temporal behavior of the reflectivities. An expansion of these exponential terms can always be done later on.

The first-order approximations (which is in fact the Born approximation) of the reflectivities are directly obtained from system (10.5) and the initial conditions (5.25):

$$r_{dpp} = \int_0^L d\zeta e^{2i\kappa_a \zeta} C',$$

$$r_{dss} = \int_0^L d\zeta e^{2i\kappa_b \zeta} H',$$

$$r_{dps} = \int_0^L d\zeta e^{i(\kappa_b + \kappa_a)\zeta} D',$$

$$r_{dsp} = P' r_{dps}.$$

$$(10.6)$$

The second-order approximations for the reflection coefficients are again obtained by substituting the formal first-order solutions of equations (5.27) and (10.5) for the quantities $\Psi_{dp}, \Psi_{ds}, t_{dps}, t_{dsp}\ \Psi_{up}, \Psi_{us}, t_{ups}$ and t_{usp} into equations (10.5) for the searched-for reflection coefficients:

$$
\begin{aligned}
r_{dpp} &= \int_0^L d\zeta \{ C(\zeta) e^{2i\kappa_a\zeta + 2\int_0^\zeta d\zeta_1 A'(\zeta_1)} \\
&\quad + 2P'D'(\zeta) e^{i(\kappa_a+\kappa_b)\zeta} \int_0^\zeta d\zeta_1 B'(\zeta_1) e^{i(\kappa_a-\kappa_b)\zeta_1} \}, \\
r_{dss} &= \int_0^L d\zeta \{ H(\zeta) e^{2i\kappa_b\zeta + 2\int_0^\zeta d\zeta_1 F'(\zeta_1)} \\
&\quad + 2P'D'(\zeta) e^{i(\kappa_a+\kappa_b)\zeta} \int_0^\zeta d\zeta_1 B'(\zeta_1) e^{i(\kappa_b-\kappa_a)\zeta_1} \}, \\
r_{dps} &= \int_0^L d\zeta \{ D(\zeta) e^{i(\kappa_b+\kappa_a)\zeta + 2\int_0^\zeta d\zeta_1 (A'(\zeta_1)+F'(\zeta_1))} \\
&\quad + C'(\zeta) e^{2i\kappa_a\zeta} \int_0^\zeta d\zeta_1 B'(\zeta_1) e^{i(\kappa_b-\kappa_a)\zeta_1} \\
&\quad + H'(\zeta) e^{2i\kappa_b\zeta} \int_0^\zeta d\zeta_1 B'(\zeta_1) e^{i(\kappa_a-\kappa_b)\zeta_1} \}, \\
r_{dsp} &= P' r_{dps},
\end{aligned}
\tag{10.7}
$$

where we have again not expanded the exponential factors of the terms of the order $O(\varepsilon)$. Also two new abbreviations have been introduced:

$$
A' = -\frac{i\omega}{Xa}\varepsilon_a, \qquad F' = -\frac{i\omega}{Yb}\varepsilon_b.
\tag{10.8}
$$

The validity of the formulas (10.7) is not restricted by the assumption of stationarity of the medium fluctuations. These formulas are applied to deterministic models.

10.3 Dynamic-Equivalent-Medium Reflectivity

Now we apply the ideas of Section 10.1 to an obliquely incident elastic plane wave onto a 1-D inhomogeneous structure. Reflectivity approximations (10.6)-(10.7) can be significantly improved using in these equations the dynamic-equivalent-medium wavenumbers instead of the wavenumbers of the homogeneous reference medium. Therefore, in the case of stationary fluctuations we perform the following substitutions

$$\kappa_a \longrightarrow \psi_P + i\gamma_P,$$
$$\kappa_b \longrightarrow \psi_{SV} + i\gamma_{SV}. \tag{10.9}$$

In the case of non-stationary fluctuations the following substitution must be performed:

$$\kappa_a \zeta \longrightarrow \kappa_a \zeta + \langle \Psi_{dp}(\zeta) \rangle,$$
$$\kappa_b \zeta \longrightarrow \kappa_b \zeta + \langle \Psi_{ds}(\zeta) \rangle. \tag{10.10}$$

Such approximations of the reflectivities we call *ODA-consistent*. Figure 61 shows at the top the exact seismogram of the P-P reflected wave compared with the Born approximation of the first order, and at the bottom the exact seismogram of the P-P reflected wave compared with the first-order ODA-consistent approximation. The incident wavelet was a causal Berlage-pulse

$$t \exp(-\alpha t) \sin(2\pi f t) \tag{10.11}$$

with $\alpha = 120s^{-1}$ and $f = 40Hz$. For the computation we took the first 100 meters of the stationary model of Figure 10. The angle of incidence ϑ was equal to $30°$. It can be observed that even for such a very complicated model (with an extremely high variance) both approximations provide satisfactory results. However, the ODA-consistent approximation is preferable (see amplitudes near the time values $35 \div 60ms$).

An even more significant improvement can be seen after considering the second-order Born approximation (10.7) of the reflectivity (see Figure 62, top). Performing now substitutions (10.9) in formula (10.7) we account for the propagation of the primary reflections and the first-order multiples (given by the second-order Born approximation) in the dynamic-equivalent medium. This (second-order ODA-consistent) approximation is displayed in Figure 62 at the bottom. It is more than a satisfactory approximation of the reflectivity for most practical needs.

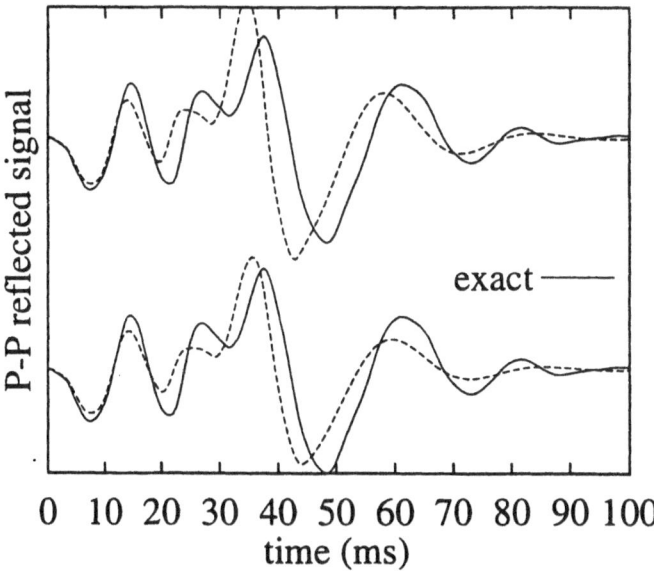

Fig. 61. The first-order Born and first-order ODA-consistent reflectivity approximations compared with the exact P-P reflectivity for $\vartheta = 30°$. The seismograms were computed for a Berlage pulse. For the stratified target we used the first 100m of the medium described in Figure 10, which is embedded between two homogeneous reference-medium half-spaces. (Reprinted from Shapiro et al., 1996, with the kind permission of Blackwell Science LTD.)

10.4 Intensity of the Reflectivity

Using the above first-order approximation of the reflectivity (10.6) we can formulate the second-order approximation for the averaged intensity of the P-P reflectivity $R_{pp}(\omega) = \left\langle r_{dpp}(\omega) r_{dpp}^*(\omega) \right\rangle$. Assuming stationarity of the random fluctuations we obtain

$$R_{pp}(\omega) = 2\frac{\omega^2}{X^2 a^2} \int_0^L dz \int_0^L dz' B_{CC}(\zeta) e^{-2i\kappa_a \zeta}, \qquad (10.12)$$

where $\zeta = z' - z$ and

$$B_{CC}(\zeta) = \langle (\varepsilon_a(0) + C_2\varepsilon_b(0) + C_1\varepsilon_\varrho(0)) \, (\varepsilon_a(\zeta) + C_2\varepsilon_b(\zeta) + C_1\varepsilon_\varrho(\zeta)) \rangle \qquad (10.13)$$

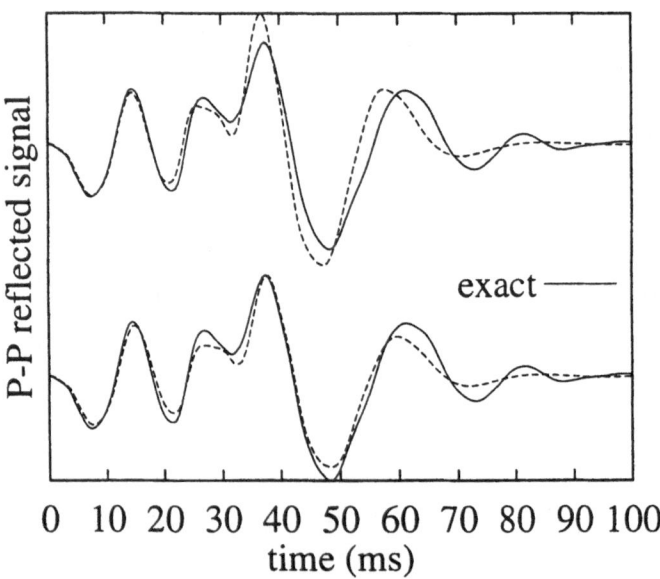

Fig. 62. Second-order reflectivity approximations compared with the exact reflectivity. The model is the same as in Figure 61. (Reprinted from Shapiro et al., 1996, with the kind permission of Blackwell Science LTD.)

is a combination of autocorrelation functions of the medium fluctuations. Introducing now a new set of integration variables ζ and $\eta = (z' + z)/2$ we obtain

$$R_{pp}(\omega) = 2\frac{\omega^2}{X^2 a^2} \int_0^L d\zeta (L - \zeta) B_{CC}(\zeta) \cos(2\kappa_a \zeta). \qquad (10.14)$$

Assuming finally that $L \gg l$ we arrive at the following approximation:

$$R_{pp}(\omega) = 2\frac{\omega^2 L}{X^2 a^2} \int_0^\infty d\zeta B_{CC} \cos(2\kappa_a \zeta). \qquad (10.15)$$

This describes the frequency- and angle dependencies of the squared absolute value (i.e, intensity) of the P-P reflectivity and can be viewed as a small-contrast approximation of the statistically extended Zöppritz equation, which describes now not the reflection coefficient for a single interface between two half spaces, but rather the reflection coefficient of a random stationary lamination embedded **between two identical half spaces.**

As an illustration of this very simple approximation, we calculated the intensity of the P-P-reflectivity of the thinly layered stack described above. As the validity condition of the above approximation is $|R_{pp}| \ll 1$ we took for the modeling the first 100 meters of the inhomogeneous medium given in Figure 10. Figure 63 compares the analytical and numerical results for $\vartheta = 30°$. The numerical results are shown for a single realization of a randomly stratified target. Approximation (10.15) predicts well the smoothed frequency-dependent reflected intensity.

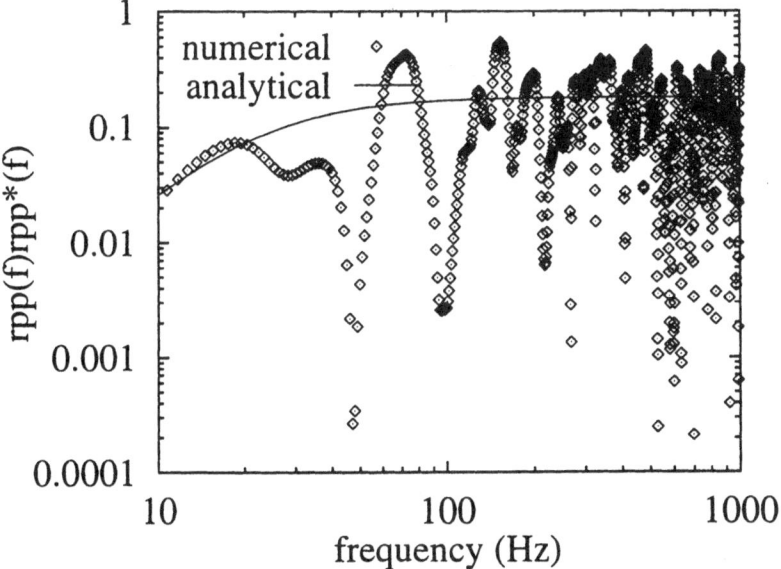

Fig. 63. Analytic approximation of the intensity of the P-P reflectivity comparison with exact one. The model is the same as for Figure 61 (Reprinted from Shapiro et al., 1996, with the kind permission of Blackwell Science LTD.)

11 Instead of Conclusions

We hope that this book will essentially serve as a contribution to solving the following practical problems always encountered in seismic exploration and in some other disciplines, like e.g., non-destructive testing:

How does one take into account the influence of small- (or even multi-) scale heterogeneities of reflector overburdens?

Our recipe is the following:

(i) In addition to a usual macro-model describing the large-scale (i.e., scale of the traveldistance to a target) heterogeneities of the medium find the *statistical macro model* of the small-scale heterogeneities.

(ii) Using both macro models improve the formulation of the 1-way or 2-way transmissivity, for example, by combining the ray-theoretical formalism with the respective generalized O'Doherty-Anstey formulas. This will improve the Green's functions.

(iii) Then, use the improved Green's functions in your modeling, imaging and inversion.

We hope that after a certainly not-easy reading of this book most readers will have gained some more intuition or deeper understanding on the above subject.

References

Aki, K., and Richards, P. G., 1980, Quantitative seismology: theory and methods, 1, W. H. Freeman and Company.

Anderson, P. W., 1958, Absence of diffusion in certain random lattices: Physical Review, 109, 1492-1505.

Arts, R. J., and Rasolofosaon, P. N. J., 1992, Complete elasticity tensor in dry and saturated rocks: Experiments versus theory: Ann. Internat. Mtg., Society of Exploration Geophysicists.

Asch, M., Kohler, W., Papanicolaou, G., Postel, M., and White, B., 1991, Frequency content of randomly scattered signals: SIAM Review, 33, 519-625.

Backus, G. E., 1962, Long-wave elastic anisotropy produced by horizontal layering: Journal of Geophysical Research, 67, 4427-4440.

Banik, N. C., Lerche, I., and Shuey, R. T., 1985, Stratigraphic filtering, part I: derivation of the O'Doherty-Anstey formula: Geophysics, 50, 2768-2774.

Bath, M., 1974, Spectral analysis in geophysics, Elsevier.

Beltzer, A. I., 1989, The effective dynamic response of random composites and polycristals - a survey of the causal approach: Wave Motion, 11, 211-229.

Bendat, J., and Piersol, A., 1986, Random data, analysis and measurement procedures, John Wiley and Sons.

Berlyand, L., and Burridge, R., 1995, The accuracy of the O'Doherty-Anstey approximation for wave propagation in highly disordered stratified media: Wave Motion, 21, 357-373.

Berryman, J. G., 1979, Long-wave elastic anisotropy in transversely isotropic media: Geophysics, 44, 896-917.

Berryman, J. G., 1986, Elastic wave attenuation in rocks containing fluids: Appl. Phys. Lett., 49, 552-554.

Biot, M. A., 1956a, Theory of propagation of elastic waves in a fluid-saturated porous solid. I. Low-frequency range: Journal of the Acoustical Society of America, 28, 168-178.

Biot, M. A., 1956b, Theory of propagation of elastic waves in a fluid-saturated porous solid. II. Higher frequency range: Journal of the Acoustical Society of America, 28, 179-191.

Biot, M. A., 1962, Generalized theory of acoustic propagation in porous dissipative media: Journal of the Acoustical Society of America, 34, 1254-1264.

Biot, M. A., 1962, Mechanics of deformation and acoustic propagation in porous media: Journal of Applied Physics, 33, 1482-1498.

Brekhovskikh, L. M., and Godin, O. A., 1990, Acoustic of layered media I. Plane and quasi-plane waves, Springer-Verlag.

Brown, R. J. S., and Korringa, J, 1975, On the dependence of the elastic properties of a porous rock on the compressibility of the pore fluid: Geophysics, 40, 608-616.

Burridge, R., and Chang, H. W., 1989, Multimode, one-dimensional wave propagation in a highly discontinuous medium: Wave Motion, 11, 231-249.

Burridge, R., de Hoop, M. V., Hsu, K., Le, L., and Norris, A., 1993, Waves in stratified viscoelastic media with microstructure: Journal of the Acoustical Society of America, 94, 2884-2894.

Burridge, R., Papanicolaou, G., and White, B. S., 1988, One-dimensional wave propagation in a highly discontinuous medium: Wave Motion, 10, 19-44.

Castagna, J., 1993, Avo analysis – tutorial and review, in Castagna, J., and Backus, M., Eds., Offset-dependent reflectivity – Theory and practice of AVO analysis, Soc. Expl. Geophys., 3–36.

Chernov, L. A., 1960, Wave propagation in a random medium, Dover Publications.

Crisanti, A., Paladin, G., and Vulpiani, A., 1993, Products of random matrices, Springer-Verlag.

de Hoop, M. V., Burridge, R., and Chang, H.-W., 1991a, Wave propagation with tunneling in a highly discontinuous layered medium: Wave Motion, 13, 307–327.

de Hoop, M. V., Chang, H.-W., and Burridge, R., 1991b, The pseudo-primary field due to a point source in a finely layered medium: Geophysical Journal International, 104, 489–506.

Dvorkin, J., Nolen-Hoeksema, R., and Nur, A., 1994, The squirt-flow mechanism: Macroscopic description: Geophysics, 59, 428–438.

Fuchs, K., and Müller, G., 1971, Computation of synthetic seismograms with the reflectivity method and comparison with observation: Geophysical Journal of the Royal Astronomical Society, 23, 417–433.

Gelinsky, S., and Shapiro, S. A., 1996, Anisotropic permeability: influence on seismic velocity and attenuation, in Fjaer, E., Holt, R. M., and Rathor, J. S., Eds., SEG Special Volume on Seismic Anisotropy, SEG, 433–461.

Gelinsky, S., and Shapiro, S. A., 1997, Dynamic-equivalent medium approach for thinly layered saturated sediments: Geophysical Journal International, 128, F1–F4.

Gelinsky, S., and Shapiro, S. A., 1997, Poroelastic Backus-averaging for anisotropic, layered fluid and gas saturated sediments: Geophysics, 62, 1867–1878.

Gilbert, F., and Backus, G.M., 1966, Propagator matrices in elastic wave and vibration problems: Geophysics, 31, 326–332.

Görich, U., and Müller, G., 1987, Apparent and intrinsic Q: the one–dimensional case: Journal of Geophysics, 61, 46–54.

Goupillaud, P. L., 1961, An approach to inverse filtering of near surface layer effects from seismic records: Geophysics, 26, 754–760.

Gredeskul, S. A., and Freilikher, V. D., 1990, Localization and wave propagation in randomly layered media: Soviet Physics Uspehi, 33, 134–146.

Gurevich, B., and Lopatnikov, S., 1995, Velocity and attenuation of elastic waves in finely layered porous rocks: Geophysical Journal International, 121, 933–947.

Hsu, K., and Burridge, R., 1991, Effects of averaging and sampling on the statistics of reflection coefficients: Geophysics, 56, 50–58.

Hubral, P., Treitel, S., and Gutowski, P., 1980, A sum autoregressive formula for the reflection response: Geophysics, 45, 1697–1705.

Ishimaru, A., 1978, Wave Propagation and Scattering in Random Media, Academic Press.

Jeffryes, B. P., 1993, Effective wave number for transmission of linear waves in one-dimensional media: Physical Review Letters, 71, 1119–1123.

Johnston, D. H., and Toksöz, M. N., 1981, Definitions and terminology, in Toksöz, M. N., and Johnston, D. H., Eds., Seismic wave attenuation, Society of Exploration Geophysicists: Geophysics reprint series, chapter I, 1–5.

Kamke, E., 1959, Differentialgleichungen. Lösungsmethoden und Lösungen. I. Gewöhnliche Differentialgleichungen, Leipzig, DDR.

Kennett, B. L. N., 1983, Seismic wave propogation in stratified media, Cambridge University Press.

Kerner, C., 1990, Modelling of soft sediments and liquid-solid interfaces: Modified wavenumber summation method and application: Geophysical Prospecting, **38**, 111–137.

Klyatskin, V. I., 1980, Stochastic equations and waves in randomly inhomogeneous media, Nauka, in Russian.

Klyatskin, V. I., 1986, The embedding method in the theory of wave propagation, Nauka, in Russian.

Klyatskin, V. I., and Saichev, A. I., 1992, Statistical and dynamical localization of plane waves in randomly layered media: Sov. Phys. Usp, **36**, 231–247.

Knoth, O., 1996, Transmissionsverluste und Amplitudenkorrektur in dünngeschichteten Medien mit inhomogener Statistik: Diploma thesis, Universität Karlsruhe.

Knoth, O., Müller, Th., and Widmaier, M., 1996, Application of the generalized O'Doherty-Anstey formula for media with inhomogeneous statistics: Ann. Internat. Mtg., European Association of Geoscientists and Engineers, Extended Abstracts, C040.

Kohler, W., Papanicolaou, G., and White, B., 1996, Localization and mode conversion for elastic waves in randomly layered media I: Wave Motion, **23**, 1–22.

Kohler, W., Papanicolaou, G., and White, B., 1996, Localization and mode conversion for elastic waves in randomly layered media II: Wave Motion, **23**, 181–201.

Korn, G. A., and Korn, T. M., 1961, Mathematical handbook for scientists and engineers. Definitions, theorems and formulas for reference and review, McGRAW-HILL BOOK COMPANY, INC.

Korn, M., 1985, Kombination von Integraltransformationen und Finite-Differenzen-Methoden zur Berechnung der Wellenausbreitung in geschichteten Medien: Ph.D. thesis, J.W. v. Goethe Universität, Frankfurt.

Lewicki, P., 1994, Long time evolution of wavefronts in random media: SIAM J. Appl. Math, **54**, 907–934.

Lewicki, P., and Burridge, R., 1996, Reflection from a deep interface in a strongly heterogeneous layered medium: Geophysical Prospecting, **44**, 571–581.

Lewicki, P., Burridge, R., and Papanicolaou, G., 1994, Pulse stabilization in a strongly heterogeneous layered medium: Wave Motion, **20**, 177–195.

Lewicki, P., and Papanicolaou, G., 1994, Reflection of wavefronts by randomly layered media: Wave Motion, **20**, 245–260.

Lifshits, I. M., Gredeskul, S. A., and Pastur, L. A., 1988, Introduction to the theory of disordered systems, John Wiley & Sons.

Menke, W., and Chen, R., 1984, Numerical studies of the coda falloff rate of multiply scattered waves in randomly layered media: Bulletin of the Seismological Society of America, **74**, 1605–1621.

Murphy III, W. F., Winkler, K. W., and Kleinberg, R. L., 1986, Acoustic relaxation in sedimentary rocks: Dependence on grain contacts and fluid saturation: Geophysics, **51**, 757–766.

Norris, A., 1993, Low-frequency dispersion and attenuation in partially saturated rocks: Journal of the Acoustical Society of America, **94**, 359–370.

Norris, A. N., 1995, Invariant embedding for elasticity and the dynamic effective medium theory: a manuscript.

O'Doherty, R. F., and Anstey, N. A., 1971, Reflections on amplitudes: Geophysical Prospecting, **19**, 430–458.

Prudnikov, A. P., Brychkov, Y. A., and Marichev, O. I., 1988, Integrals and series, Gordon and Breach Science Publ.

Resnick, J. R., Lerche, I., and Shuey, R. T., 1986, Reflection, transmission and the generalized primary wave: Geophysical Journal of the Royal Astronomical Society, **87**, 349–377.

Robinson, E. A., 1967, Multichannel time series analysis with digital computer programs, Holden-Day, Inc., San Francisco.

Robinson, E. A., 1975, Dynamic predictive deconvolution: Geophysical Prospecting, 779–797.

Robinson, E. A., and Treitel, S., 1977, The spectral function of a layered system and the determination of the waveforms at depth: Geophysical Prospecting, 434–459.

Robinson, E. A., and Treitel, S., 1980, Geophysical signal analysis, Prentice-Hall, Inc., Englewood Cliffs., N.J.

Rytov, S. M., Kravtsov, Yu. A., and Tatarskii, V. I., 1989a, Elements of random process theory, Principles of statistical radiophysics, **1**, Springer Verlag.

Rytov, S. M., Kravtsov, Yu. A., and Tatarskii, V. I., 1989b, Wave propagation through random media, Principles of statistical radiophysics, **4**, Springer Verlag.

Sato, H., 1995, Formulation of the multiple non-isotropic scattering process in 3-D space on the basis of energy transport theory: Geophysical Journal International, **121**, 523–531.

Schmitt, D. P., 1989, Acoustic multipole logging in transversely isotropic poroelastic formations: Journal of the Acoustical Society of America, **86**, 2397–2421.

Schoenberg, M., and Muir, F., 1989, A calculus for finely layered anisotropic media: Geophysics, **54**, 581–589.

Schoenberger, M., and Levin, F. K., 1974, Apparent attenuation due to intrabed multiples: Geophysics, **39**, 278–291.

Shapiro, S. A., and Hubral, P., 1994, Generalized O'Doherty–Anstey formula for P-SV waves in random multilayered elastic media: 64th Meeting, SEG, Expanded Abstracts., Expanded Abstracts, 1426-1429.

Shapiro, S. A., and Hubral, P., 1995, Frequency-dependent shear-wave splitting and velocity anisotropy due to elastic multilayering: Journal of Seismic Exploration, **4**, 151–168.

Shapiro, S. A., and Hubral, P., 1996, Elastic waves in thinly layered sediments: The equivalent medium and generalized O'Doherty-Anstey formulas: Geophysics, **61**, 1282–1300.

Shapiro, S. A., Hubral, P., and Ursin, B., 1996, Reflectivity/transmissivity for 1-d inhomogeneous random elastic media: dynamic-equivalent-medium approach: Geophysical Journal International, **126**, 184–196.

Shapiro, S. A., Hubral, P., and Zien, H., 1994b, Frequency-dependent anisotropy of scalar waves in a multilayered medium: Journal of Seismic Exploration, **3**, 37–52.

Shapiro, S. A., and Kneib, G., 1993, Seismic attenuation by scattering: Theory and numerical results: Geophysical Journal International, **114**, 373–391.

Shapiro, S. A., and Müller, T., 1997, Seismic signatures of disorder and transport properties in porous rocks: Ann. Internat. Mtg., Society of Exploration Geophysicists, Expanded abstracts, 1005–1008.

Shapiro, S. A., Schwarz, R., and Gold, N., 1996, The effect of random isotropic inhomogeneities on the phase velocities of seismic waves: Geophysical Journal International, **127**, 783–794.

Shapiro, S. A., and Treitel, S., 1997, Multiple scattering of seismic waves in multilayered structures: Physics of the Earth and Planetary Interiors, **104**, 147–159.

Shapiro, S. A., and Zien, H., 1993, The O'Doherty–Anstey formula and localization of seismic waves: Geophysics, 736–740.

Shapiro, S. A., Zien, H., and Hubral, P., 1994a, A generalized O'Doherty–Anstey formula for waves in finely–layered media: Geophysics, **59**, 1750–1762.

Sheng, P., 1995, Introduction to Wave Scattering, Localization, and Mesoscopic Phenomena, Academic Press.

Sheng, P., White, B., Zhang, Z.-Q., and Papanicolaou, G., 1990, Wave localization and multiple scattering in randomly–layered media, *in* Sheng, P., Ed., Scattering and localization of classical waves in random media, World Scientific, 563–619.

Sipe, J. E., Sheng, P., White, B. S., and Cohen, M. H., 1988, Brewster anomalies: a polarization-induced delocalization effect: Phys. Rev. Lett, **60**, 108–111.

Stanke, F. E., and Burridge, R., 1993, Spatial versus ensemble averaging for modeling wave propagation in finely layered media: Journal of the Acoustical Society of America, **93**, 36–41.

Tygel, M., and Hubral, P., 1987, Transient waves in layered media, Elsevier.

Ursin, B., 1983, Review of elastic and electromagnetic wave propagation in horizontally layered media: Geophysics, **48**, 1063–1081.

Ursin, B., 1987, The plane–wave reflection and transmission response of a vertically inhomogeneous acoustic medium: *in* Bernabini, M., Carrion, P., Jacovitti, G., Rocca, F., Treitel, S., and Worthington, M., Eds., Deconvolution and inversion, 189–207, Oxford. Blackwell Scientific Publications, Proceedings of a workshop in Rome, 3–5 September.

Ursin, B., 1990, Offset-dependent geometrical spreading in a layered medium: Geophysics, **55**, 492–496.

Wapenaar, C. P. A., Slot, R. E., and Herrmann, F. J., 1994, Towards an extended macro model, that takes fine-layering into account: Journal of Seismic Exploration, **3**, 245–260.

Werner, U., and Shapiro, S. A., 1997, Frequency-dependent shear wave splitting due to transversly-isotropic multilayering: Ann. Internat. Mtg., Society of Exploration Geophysicists, Expanded abstracts, 1858–1861.

Werner, U., and Shapiro, S. A., 1998, Intrinsic anisotropy and thin multilayering - two anisotropy effects combined: Geophysical Journal International, **132**, 363–373.

White, B., Sheng, P., and Nair, B., 1990, Localization and backscattering spectrum of seismic waves in stratified lithology: Geophysics, **55**, 1158–1165.

White, J. E., 1983, Underground Sound. Application of Seismic Waves, Elsevier.

Widmaier, M., 1996, Amplitude-preserving migration and AVO analysis corrected for thin layering: Ph.D. thesis, Universität Karlsruhe.

Widmaier, M., Müller, Th., Shapiro, S. A., and Hubral, P., 1995, Amplitude-preserving migration and elastic P-wave AVO corrected for thin layering: Journal of Seismic Exploration, **4**, 169–177.

Widmaier, M., Shapiro, S. A., and Hubral, P., 1996, AVO correction for a thinly layered reflector overburden: Geophysics, **61**, 520–528.

Winterstein, D., 1990, Velocity anisotropy terminology for geophysicists: Geophysics, **55**, 1070–1088.

Wrolstad, K., 1993, Offset-Dependent Amplitude Analysis of Data from the Veslefrikk Field, Offshore Norway, *in* Castagna, J., and Backus, M., Eds., Offset-dependent reflectivity – Theory and practice of AVO analysis, SEG, 250–266.

Zien, H., 1993, The effect of micro-layering on the wave propagation: Ph.D. thesis, University of Karlsruhe.

Index

acoustic impedance, 25, 38
acoustic waves, 47
amplitude-variation-with-offset analysis, 119
anisotropy, 9, 93, 98
attenuation, 9, 15, 77, 87, 93
attenuation coefficient, 9, 13, 14, 17, 46, 50-52, 63, 73, 87, 89, 91, 97, 103, 112
autocorrelation function, 11, 38, 57
averaged velocity, 53
averaging, 9, 49, 51, 78, 90

Backus averaging, 6, 61, 87, 89, 92, 100
Born approximation, 70
Brewster anomaly effect, 57

central-limit theorem, 19
coda, 24, 33, 36, 109
common-offset gather, 122
correlation lag, 11, 53, 58
correlation length, 12, 15, 23, 34, 87
crosscorrelation, 57, 75

Darcy-coefficient, 155
deconvolution, 23, 169
density log, 12
depth domain, 23
dispersion, 87
dynamic predictive deconvolution, 171
dynamic-equivalent medium, 4, 51, 61, 73, 74, 87, 91, 98

effective transversely-isotropic elastic medium, 89
eigenvalue decomposition, 65
elastic media, 61
elastic plane waves, 61
elastic scattering, 156
elasticity theory, 1
ensemble, 13
ensemble averaging, 13, 16

ergodic random process (medium), 13, 14, 34, 90
exponential correlation function, 79, 89, 91
exponential medium, 12

Fürstenberg's theorem, 19
finite-difference, 6
first arrival, 32, 36
first statistical moment, 10
fluctuation, 10, 12, 50, 53, 64, 70, 71, 87, 95
fluctuation spectrum, 11
Fourier transform, 11, 89
fourth statistical moments, 24, 35, 37
fractal, 13, 139, 163
fractal dimension, 13
frequency-dependent shear-wave splitting, 77, 87, 99
frequency-dependent velocity anisotropy, 87, 93
fundamental solution, 48
fundamental polynomials, 25, 26

Gaussian autocorrelation functions, 12
Gaussian medium, 12
Gaussian pulse, 112
generalized O'Doherty-Anstey formulas, 4, 6, 18, 24, 45, 70
generalized primary, 24, 33, 36, 109
geometrical optics, 87
global flow, 156
Goupillaud model, 23, 24, 38, 169
Green's functions, 181

homogeneous reference medium, 88
horizontal slowness, 53
horizontal wavenumber, 16

inelasticity, 14
initial conditions, 28, 70

interlayer flow, 156
intrinsic anisotropy, 102
intrinsic attenuation, 14
invariant-embedding method, 6, 61
inversion, 23

Kirchhoff migration, 122
Kramers-Krönig dispersion relation,
    73

length of the generalized primary, 36
linear filtering, 1
linear-system theory, 1
localization, 9, 14, 18, 70
localization length, 14, 20
Lyapunov characteristic exponent, 19

mathematical expectation, 10
mean value, 10
meanfield, 15
meanfield attenuation, 15
minimum phase, 73
mode conversion, 61, 111
multilayered, 1
multimode wave propagation, 49, 61
multiple scattering, 9, 14, 23, 25, 27,
    31
multiplicative ergodic theorem, 19

O'Doherty-Anstey formulas, 4, 6, 18,
    24, 45, 70
Oseledets's theorem, 20

permeability, 152, 161
phase reconstruction, 96
phase velocity, 46, 73, 90, 96
plane wave, 16, 23
poroelastic medium, 152
poroelasticity, 151
power spectrum, 11
power spectrum of the reflection-coefficient
    series, 54
pressure wave, 14, 23, 53
probability densities, 9
propagator matrix, 6, 19, 20, 67, 94
pulse stabilization, 20, 115

quality factor, 93

random fields, 9
random function, 9
random matrices, 7, 18, 48
random medium, 9, 10, 13, 23, 91,
    96, 103
random process, 9
ray theory, 90
Rayleigh scattering, 89, 163
realizations of a random medium, 13
reciprocal quality factor, 88, 93
reflection and transmission coefficients,
    24
reflection-coefficient series, 32, 34, 38,
    54
reflectivity, 3, 24, 25, 27–29, 62, 68,
    169
reflectivity method, 6
relative standard deviation, 104
Rytov approximation, 70

scalar waves, 23, 45, 52
second statistical moment, 11, 24, 37
self averaging, 7, 17, 18, 20, 115
self-averaged quantities, 9, 14, 16,
    103
SH-wave, 45, 52, 53
shear-wave birefringence, 99
shear-wave splitting, 92, 99, 101
single-scattering approximation, 31
small-perturbation expansion, 61
small-scale heterogeneities, 1
sonic log, 12
spatial averaging, 14
stack of thin layers, 1
standard deviation, 11, 12
stationary Gaussian random process,
    35
stationary medium, 87
stationary process, 10
stationary-in-the-wide-sense, 10
statistic analysis, 9
statistical inversion, 105
statistical macro model, 4, 80, 181
statistical moments, 10

statistically homogeneous, 10
stratified, 1
stratified layers, 1
stratified structures, 1
stratigraphic filter, 2
stratigraphic filtering, 1, 16, 87, 133
strong localization, 14
symplectic group, 20
systems with disorder, 17

time domain, 23
time-harmonic transmissivity, 2, 52,
        74, 87
transient transmissivity, 2, 37, 109
transmissivity, 2, 16, 24, 25, 27–29,
        32, 36, 45, 49, 52, 56, 61,
        62, 68, 70, 77, 92, 93, 119,
        120, 155, 181
transmitted wavefield, 2
transverse isotropy, 9
traveltime, 50
turbulence fluctuations, 13
two-way normal incidence travel time,
        24
two-way stratigraphic filtering, 119
two-way transmissivities, 32
typical realization, 16

unwrapped phase, 105

variance, 11, 35
velocity dispersion, 9
vertical-phase increment, 4, 16, 17,
        50–52, 63, 87, 91, 103
Virtser's theorem, 19, 20
von Karman correlation function, 13,
        163

wavefield localization, 7
wavefields, 2
wavelength, 87
weak localization, 15, 70

z-transforms, 25

Lecture Notes in Earth Sciences

For information about Vols. 1–19
please contact your bookseller or Springer-Verlag

Vol. 20: P. Baccini (Ed.), The Landfill. IX, 439 pages. 1989.

Vol. 21: U. Förstner, Contaminated Sediments. V, 157 pages. 1989.

Vol. 22: I. I. Mueller, S. Zerbini (Eds.), The Interdisciplinary Role of Space Geodesy. XV, 300 pages. 1989.

Vol. 23: K. B. Föllmi, Evolution of the Mid-Cretaceous Triad. VII, 153 pages. 1989.

Vol. 24: B. Knipping, Basalt Intrusions in Evaporites. VI, 132 pages. 1989.

Vol. 25: F. Sansò, R. Rummel (Eds.), Theory of Satellite Geodesy and Gravity Field Theory. XII, 491 pages. 1989.

Vol. 26: R. D. Stoll, Sediment Acoustics. V, 155 pages. 1989.

Vol. 27: G.-P. Merkler, H. Militzer, H. Hötzl, H. Armbruster, J. Brauns (Eds.), Detection of Subsurface Flow Phenomena. IX, 514 pages. 1989.

Vol. 28: V. Mosbrugger, The Tree Habit in Land Plants. V, 161 pages. 1990.

Vol. 29: F. K. Brunner, C. Rizos (Eds.), Developments in Four-Dimensional Geodesy. X, 264 pages. 1990.

Vol. 30: E. G. Kauffman, O.H. Walliser (Eds.), Extinction Events in Earth History. VI, 432 pages. 1990.

Vol. 31: K.-R. Koch, Bayesian Inference with Geodetic Applications. IX, 198 pages. 1990.

Vol. 32: B. Lehmann, Metallogeny of Tin. VIII, 211 pages. 1990.

Vol. 33: B. Allard, H. Borén, A. Grimvall (Eds.), Humic Substances in the Aquatic and Terrestrial Environment. VIII, 514 pages. 1991.

Vol. 34: R. Stein, Accumulation of Organic Carbon in Marine Sediments. XIII, 217 pages. 1991.

Vol. 35: L. Håkanson, Ecometric and Dynamic Modelling. VI, 158 pages. 1991.

Vol. 36: D. Shangguan, Cellular Growth of Crystals. XV, 209 pages. 1991.

Vol. 37: A. Armanini, G. Di Silvio (Eds.), Fluvial Hydraulics of Mountain Regions. X, 468 pages. 1991.

Vol. 38: W. Smykatz-Kloss, S. St. J. Warne, Thermal Analysis in the Geosciences. XII, 379 pages. 1991.

Vol. 39: S.-E. Hjelt, Pragmatic Inversion of Geophysical Data. IX, 262 pages. 1992.

Vol. 40: S. W. Petters, Regional Geology of Africa. XXIII, 722 pages. 1991.

Vol. 41: R. Pflug, J. W. Harbaugh (Eds.), Computer Graphics in Geology. XVII, 298 pages. 1992.

Vol. 42: A. Cendrero, G. Lüttig, F. Chr. Wolff (Eds.), Planning the Use of the Earth's Surface. IX, 556 pages. 1992.

Vol. 43: N. Clauer, S. Chaudhuri (Eds.), Isotopic Signatures and Sedimentary Records. VIII, 529 pages. 1992.

Vol. 44: D. A. Edwards, Turbidity Currents: Dynamics, Deposits and Reversals. XIII, 175 pages. 1993.

Vol. 45: A. G. Herrmann, B. Knipping, Waste Disposal and Evaporites. XII, 193 pages. 1993.

Vol. 46: G. Galli, Temporal and Spatial Patterns in Carbonate Platforms. IX, 325 pages. 1993.

Vol. 47: R. L. Littke, Deposition, Diagenesis and Weathering of Organic Matter-Rich Sediments. IX, 216 pages. 1993.

Vol. 48: B. R. Roberts, Water Management in Desert Environments. XVII, 337 pages. 1993.

Vol. 49: J. F. W. Negendank, B. Zolitschka (Eds.), Paleolimnology of European Maar Lakes. IX, 513 pages. 1993.

Vol. 50: R. Rummel, F. Sansò (Eds.), Satellite Altimetry in Geodesy and Oceanography. XII, 479 pages. 1993.

Vol. 51: W. Ricken, Sedimentation as a Three-Component System. XII, 211 pages. 1993.

Vol. 52: P. Ergenzinger, K.-H. Schmidt (Eds.), Dynamics and Geomorphology of Mountain Rivers. VIII, 326 pages. 1994.

Vol. 53: F. Scherbaum, Basic Concepts in Digital Signal Processing for Seismologists. X, 158 pages. 1994.

Vol. 54: J. J. P. Zijlstra, The Sedimentology of Chalk. IX, 194 pages. 1995.

Vol. 55: J. A. Scales, Theory of Seismic Imaging. XV, 291 pages. 1995.

Vol. 56: D. Müller, D. I. Groves, Potassic Igneous Rocks and Associated Gold-Copper Mineralization. 2nd updated and enlarged Edition. XIII, 238 pages. 1997.

Vol. 57: E. Lallier-Vergès, N.-P. Tribovillard, P. Bertrand (Eds.), Organic Matter Accumulation. VIII, 187 pages. 1995.

Vol. 58: G. Sarwar, G. M. Friedman, Post-Devonian Sediment Cover over New York State. VIII, 113 pages. 1995.

Vol. 59: A. C. Kibblewhite, C. Y. Wu, Wave Interactions As a Seismo-acoustic Source. XIX, 313 pages. 1996.

Vol. 60: A. Kleusberg, P. J. G. Teunissen (Eds.), GPS for Geodesy. VII, 407 pages. 1996.

Vol. 61: M. Breunig, Integration of Spatial Information for Geo-Information Systems. XI, 171 pages. 1996.

Vol. 62: H. V. Lyatsky, Continental-Crust Structures on the Continental Margin of Western North America. XIX, 352 pages. 1996.

Vol. 63: B. H. Jacobsen, K. Mosegaard, P. Sibani (Eds.), Inverse Methods. XVI, 341 pages, 1996.

Vol. 64: A. Armanini, M. Michiue (Eds.), Recent Developments on Debris Flows. X, 226 pages. 1997.

Vol. 65: F. Sansò, R. Rummel (Eds.), Geodetic Boundary Value Problems in View of the One Centimeter Geoid. XIX, 592 pages. 1997.

Vol. 66: H. Wilhelm, W. Zürn, H.-G. Wenzel (Eds.), Tidal Phenomena. VII, 398 pages. 1997.

Vol. 67: S. L. Webb, Silicate Melts. VIII. 74 pages. 1997.

Vol. 68: P. Stille, G. Shields, Radiogenetic Isotope Geochemistry of Sedimentary and Aquatic Systems. XI, 217 pages. 1997.

Vol. 69: S. P. Singal (Ed.), Acoustic Remote Sensing Applications. XIII, 585 pages. 1997.

Vol. 70: R. H. Charlier, C. P. De Meyer, Coastal Erosion – Response and Management. XVI, 343 pages. 1998.

Vol. 71: T. M. Will, Phase Equilibria in Metamorphic Rocks. XIV, 315 pages. 1998.

Vol. 72: J. C. Wasserman, E. V. Silva-Filho, R. Villas-Boas (Eds.), Environmental Geochemistry in the Tropics. XIV, 305 pages. 1998.

Vol. 73: Z. Martinec, Boundary-Value Problems for Gravimetric Determination of a Precise Geoid. XII, 223 pages. 1998.

Vol. 74: M. Beniston, J. L. Innes (Eds.), The Impacts of Climate Variability on Forests. XIV, 329 pages. 1998.

Vol. 75: H. Westphal, Carbonate Platform Slopes – A Record of Changing Conditions. XI, 197 pages. 1998.

Vol. 76: J. Trappe, Phanerozoic Phosphorite Depositional Systems. XII, 316 pages. 1998.

Vol. 77: C. Goltz, Fractal and Chaotic Properties of Earthquakes. XIII, 178 pages. 1998.

Vol. 78: S. Hergarten, H. J. Neugebauer (Eds.), Process Modelling and Landform Evolution. X, 305 pages. 1999.

Vol. 79: G. H. Dutton, A Hierarchical Coordinate System for Geoprocessing and Cartography. XVIII, 231 pages. 1999.

Vol. 80: S. A. Shapiro, P. Hubral, Elastic Waves in Random Media. XIV, 191 pages. 1999.

Vol. 81: Y. Song, G. Müller, Sediment-Water Interactions in Anoxic Freshwater Sediments. VI, 111 pages. 1999.

Vol. 82: T. M. Løseth, Submarine Massflow Sedimentation. IX, 156 pages. 1999.